Microbiology
PreTest® Self-Assessment and Review
Seventh Edition

Edited by

Richard C. Tilton, Ph.D.
President and Chief Scientific Officer
North American Laboratory Group
New Britain, Connecticut

Raymond W. Ryan, Ph.D.
Associate Professor
Department of Laboratory Medicine
University of Connecticut School of Medicine
Farmington, Connecticut

McGraw-Hill, Inc.
Health Professions Division/PreTest Series

New York   St. Louis   San Francisco   Auckland
Bogotá   Caracas   Lisbon   London   Madrid
Mexico   Milan   Montreal   New Delhi   Paris
San Juan   Singapore   Sydney   Tokyo   Toronto

Microbiology: PreTest® Self-Assessment and Review, Seventh Edition

International Editions 1993

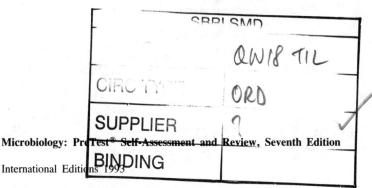

Exclusive rights by McGraw-Hill Book Co.-Singapore for manufacture and export. This book cannot be re-exported from the country to which it is consigned by McGraw-Hill.

Copyright © 1993 1991 1988 1985 1983 1980 1976 by McGraw-Hill, Inc. All rights reserved. Except as permitted under the Copyright Act of 1976, no part of this publication may be reproduced or distributed in any form or by any means, or stored in a data base or retrieval system, without the prior written permission of the publisher.

1 2 3 4 5 6 7 8 9 0 KHL 9 8 7 6 5 4 3 2

Library of Congress Cataloging-in-Publication Data

Microbiology: PreTest self-assessment and review / edited by
   Richard C. Tilton, Raymond W. Ryan.—7th ed.
     p.   cm.
   Includes bibliographical references.
   ISBN 0-07-052000-3
    1. Medical microbiology—Examinations, questions, etc.
 I. Tilton, Richard C. II. Ryan, Raymond W.
   [DNLM: 1. Microbiology—examination questions.
   QW 18 M626]
 QR61.5.M57   1993
 616'.01'076—dc20
 DNLM/DLC
 for Library of Congress   91-38958
                                                     CIP

The editors were Gail Gavert and Bruce MacGregor.
The production supervisor was Gyl A. Favours.
This book was set in Times Roman by Compset.

When ordering this title use ISBN 0-07-112978-2

Printed in Singapore

Contents

Introduction *vii*

Virology
  Questions *1*
  Answers, Explanations, and References *20*

Bacteriology
  Questions *40*
  Answers, Explanations, and References *60*

Physiology
  Questions *82*
  Answers, Explanations, and References *94*

Rickettsiae, Chlamydiae, and Mycoplasmas
  Questions *104*
  Answers, Explanations, and References *109*

Mycology
  Questions *114*
  Answers, Explanations, and References *120*

Parasitology
  Questions *127*
  Answers, Explanations, and References *141*

Immunology
  Questions *151*
  Answers, Explanations, and References *166*

Bibliography *181*

Introduction

Microbiology: PreTest® Self-Assessment and Review provides medical students, as well as physicians, with a comprehensive and convenient instrument for self-assessment and review within the field of microbiology. The 500 questions parallel the format and degree of difficulty of the questions contained in Step 1 of the United States Medical Licensing Examination (USMLE) as well as the Foreign Medical Graduate Examination in the Medical Sciences (FMGEMS).

To conform with current USMLE guidelines, all K-type (multiple true-false) questions have been eliminated.

Each question in the book is accompanied by an answer, a paragraph explanation, and a specific page reference to a current journal article, a textbook, or both. A bibliography, listing all the sources used in the book, follows the last chapter.

Perhaps the most effective way to use this book is to allow yourself one minute to answer each question in a given chapter; as you proceed, indicate your answer beside each question. By following this suggestion, you will be approximating the time limits imposed by the board examinations previously mentioned.

When you finish answering the questions in a chapter, you should then spend as much time as you need verifying your answers and carefully reading the explanations. Although you should pay special attention to the explanations for the questions you answered incorrectly, you should read *every* explanation. The authors of this book have designed the explanations to reinforce and supplement the information tested by the questions. If, after reading the explanations for a given chapter, you feel you need still more information about the material covered, you should consult and study the references indicated.

Virology

DIRECTIONS: Each question below contains five suggested responses. Select the **one best** response to each question.

1. A patient presents with a request for testing for human immunodeficiency virus (HIV) because of a weekend fling of promiscuous sexual activity 2 weeks ago. Given that time frame of possible infection, what is the best, most cost-efficient test to order?

   (A) HIV enzyme immunoassay (EIA) for antibody
   (B) HIV EIA for antigen
   (C) HIV Western blot
   (D) Polymerase chain reaction (PCR) for HIV
   (E) HIV culture

2. Rabies is an acute infection of the central nervous system. Which of the following statements best describes the rabies virus?

   (A) It has multiple antigenic types
   (B) It can be isolated from the blood of infected patients
   (C) It can be transmitted by a dog 4 weeks before the dog becomes noticeably ill
   (D) It produces an infection that is nearly always fatal to humans
   (E) It can easily be transmitted from person to person

3. All the following statements about cytomegalovirus infection are true EXCEPT

   (A) it can be cultured from the red blood cells of infected persons
   (B) it can be transmitted transplacentally
   (C) it can be activated by immunosuppressive agents
   (D) it will rarely cause clinically apparent disease
   (E) it can cause chorioretinitis

4. True statements about respiratory syncytial virus (RSV) include all the following EXCEPT that

   (A) it is the major cause of lower respiratory tract illness in young children
   (B) it is an enveloped RNA virus belonging to the Paramyxoviridae
   (C) it grows well in human heteroploid cells, forming characteristic cytopathic effects (syncytia)
   (D) its resistance to changes in temperature and pH is poor
   (E) its diagnosis is confirmed by isolation of the virus from rectal specimens, which yield the highest recovery rate

5. Epstein-Barr virus is the etiologic agent of infectious mononucleosis. It has also been implicated in Burkitt's lymphoma and nasopharyngeal carcinoma. What other virus has been shown to cause infectious mononucleosis–like disease?

(A) Herpes simplex type 1
(B) Respiratory syncytial virus
(C) Cytomegalovirus
(D) Rubella virus
(E) Adenovirus

6. Antigenic variation in influenza virus is a phenomenon that

(A) occurs by the addition of new antigenic determinants
(B) is more prominent in B strains than in A strains
(C) affects only the hemagglutinin and not the neuraminidase antigens
(D) has little impact on the clinical management of influenza
(E) occurs each year in the fall just prior to the onset of "flu" season

7. Which of the following viruses causes an acute febrile rash and produces disease in immunocompetent children, but has been associated with transient aplastic crisis in persons with sickle cell disease?

(A) Rubeola
(B) Varicella zoster
(C) Parvovirus
(D) Rubella
(E) Herpes simplex

8. Infection with herpes simplex virus, a common human pathogen, is best described by which of the following statements?

(A) The CNS and visceral organs are usually involved
(B) It rarely recurs in a host who has a high antibody titer
(C) It can be reactivated by emotional disturbances or prolonged exposure to sunlight
(D) Initial infection usually occurs by intestinal absorption of the virus
(E) Infection with type 1 virus is most common

9. All the following statements about human rotaviruses are true EXCEPT that they

(A) produce an infection that is seasonally distributed, peaking in fall and winter
(B) produce cytopathic effects in many conventional tissue culture systems
(C) are non-lipid-containing RNA viruses possessing a double-shelled capsid
(D) can be sensitively and rapidly detected in stools by the enzyme-linked immunosorbent assay (ELISA) technique
(E) have been implicated as a major etiologic agent of infantile gastroenteritis

10. All the following statements regarding herpes simplex virus are true EXCEPT

(A) it is a large virus containing double-stranded DNA
(B) it is subject to available, effective antiviral therapy
(C) it is effectively controlled by vaccination
(D) it causes latent infection, which can be reactivated
(E) it is related to cytomegalovirus, Epstein-Barr virus, and varicella-zoster virus

11. Increased replication of HIV-1 is effected by which of the following retroviral genes?

(A) *Env*
(B) *Gag*
(C) *Tat*
(D) *Pol*
(E) *Nef*

12. The Filoviridae are a newly recognized family of negative-sense, single-stranded RNA viruses. Which of the following viruses belongs to this family and causes hemorrhagic fever?

(A) Marburg virus
(B) Yellow fever virus
(C) Dengue virus
(D) Lassa fever virus
(E) Parvovirus

13. Molluscum contagiosum virus, a poxvirus, can be described by which of the following statements?

(A) It has not been transmitted experimentally to humans
(B) It differs from other poxviruses in appearance on electron microscopy
(C) It produces proliferative lesions on genital epithelium
(D) It usually infects older persons
(E) It is spread by aerosolization

14. All the following categories of people are considered at high risk for HIV infection EXCEPT

(A) sexually active homosexual males
(B) sexually active homosexual females
(C) heterosexual IV drug abusers
(D) infants born to mothers infected by HIV
(E) heterosexual sexual partners of persons with AIDS

15. Rotavirus is a common cause of diarrhea in children. Which of the following statements best describes rotavirus?

(A) It is an RNA virus
(B) Tests for detection of antigen are rarely useful
(C) Culture is the routine method of confirming infection
(D) It is rarely a nosocomial pathogen
(E) Person-to-person transmission is rare

16. Human T-lymphotropic virus type I (HTLV-I) is associated with all the following syndromes EXCEPT

(A) adult T-cell leukemia
(B) hairy-cell leukemia
(C) adult T-cell lymphoma
(D) tropical spastic paraparesis
(E) HTLV-I–associated myelopathy

17. Infectious mononucleosis, a viral disorder that can be debilitating, is characterized by which of the following statements?

(A) It is most prevalent in children less than 14 years old
(B) It is caused by a rhabdovirus
(C) The causative pathogen is an Epstein-Barr virus
(D) Affected persons respond to treatment with the production of heterophil antibodies
(E) Ribavirin is the treatment of choice

18. Coronaviruses, which have protuberances that give them an appearance like that of a solar corona, are

(A) single-stranded RNA viruses
(B) nonenveloped viruses
(C) usually isolated during the summer months
(D) assembled in the nucleus of host cells
(E) closely related to polyomaviruses

19. A tube of monkey kidney cells is inoculated with nasopharyngeal secretions. During the next 7 days, no cytopathic effects (CPEs) are observed. On the eighth day, the tissue culture is infected accidentally with a picornavirus; nevertheless, the culture does not develop CPEs. The most likely explanation of this phenomenon is that

(A) the nasopharyngeal secretions contained hemagglutinins
(B) the nasopharyngeal secretions contained rubella virus
(C) picornavirus does not produce CPEs
(D) picornavirus does not replicate in monkey kidney cells
(E) monkey kidney cells are resistant to CPEs

20. The best temperature for transport of patient specimens to the laboratory for viral culture is

(A) 42°C
(B) 37°C
(C) 25°C
(D) 4°C
(E) −20°C

21. One characteristic of arboviruses is that they

(A) are transmitted by arthropod vectors
(B) are usually resistant to ether
(C) usually cause symptomatic infection in humans
(D) have a genome of double-stranded DNA
(E) are closely related to parvoviruses

22. All the following statements describing subacute sclerosing panencephalitis (SSPE) are true EXCEPT
(A) it is a progressive disease involving both white and gray matter
(B) it is a late CNS manifestation of measles
(C) it is a rare event occurring in 1 of 300,000 cases of measles
(D) viral RNA can be demonstrated in brain cells
(E) demyelination is characteristic

23. Which of the following statements best describes interferon's suspected mode of action in producing resistance to viral infection?
(A) It stimulates cell-mediated immunity
(B) It stimulates humoral immunity
(C) Its direct antiviral action is related to the suppression of messenger RNA formation
(D) Its action is related to the synthesis of a protein that inhibits translation or transcription
(E) It alters the permeability of the cell membrane so that viruses cannot enter the cell

24. All the following viruses belong to the family Picornaviridae EXCEPT
(A) coxsackievirus group A
(B) coxsackievirus group B
(C) poliovirus
(D) rabies virus
(E) echovirus

25. Coronaviruses are recognized by club-shaped surface projections that are 20 nm long and resemble solar coronas. These viruses are characterized by their ability to
(A) infect infants more frequently than adults
(B) cause the common cold
(C) grow well in the usual cultured cell lines
(D) grow profusely at 50°C
(E) agglutinate human red blood cells

26. Although viral infections usually precede or accompany the onset of Reye's syndrome, the actual role of the viral infection is not known. Reye's syndrome is most often associated with
(A) outbreaks of influenza B
(B) primary infection with herpesvirus
(C) infection with rotavirus
(D) infection with rhinovirus
(E) respiratory syncytial virus (RSV) disease

27. Infectious mononucleosis is associated with all the following EXCEPT
(A) infection with a nonlatent virus
(B) heterophil antibodies
(C) bilateral hilar adenopathy on chest x-ray
(D) lymphadenopathy
(E) atypical lymphocytes

28. An obstetrician sees a pregnant patient who has been exposed to the rubella virus in the 18th week of gestation. She has no history of rubella and on serologic testing has no detectable antibody. The best course of action would be to

(A) vaccinate her immediately
(B) administer rubella immune serum
(C) suggest termination of pregnancy
(D) follow her antibody titer serologically
(E) assure her that no problem exists because she is past the first trimester of pregnancy

29. The ability to adsorb to receptors on red blood cells is a feature of some viruses, including the paramyxoviruses. A test to demonstrate paramyxoviral hemagglutinins should be performed at 4°C because at this temperature

(A) the cell receptor site is unmasked
(B) the activity of the hemagglutinin is preserved
(C) the agglutination of red blood cells is enhanced by cold agglutinins
(D) neuraminidase is activated
(E) the infectivity of the virus is decreased

30. Delta hepatitis only occurs in patients who also have either acute or chronic infection with hepatitis B virus. The delta agent is

(A) an incomplete hepatitis B virus
(B) related to hepatitis A virus
(C) a hepatitis B mutant
(D) an incomplete RNA virus
(E) the agent of non-A, non-B hepatitis

31. Mumps virus is biologically related to the virus causing which of the following diseases?

(A) Rabies
(B) Hepatitis A
(C) Measles
(D) Variola
(E) Varicella

32. Which of the following antiviral agents is a purine nucleoside analogue that has shown promise with Lassa fever, influenza A and B, and respiratory syncytial virus (RSV)?

(A) Amantadine
(B) Rimantadine
(C) Vidarabine
(D) Ribavirin
(E) Acyclovir

33. Echoviruses are cytopathogenic human viruses that mainly infect the

(A) respiratory system
(B) central nervous system
(C) blood and lymphatic systems
(D) intestinal tract
(E) bladder and urinary tract

Virology

34. Acute herpetic gingivostomatitis, which is the most common primary infection with type 1 herpes simplex, is a disease that

(A) has an incubation period of 2 weeks
(B) is most common in adolescents
(C) usually does not cause fever
(D) causes regional lymphadenitis
(E) does not recur

35. Acute hemorrhagic conjunctivitis (AHC) is a contagious ocular infection characterized by pain, swelling of the eyelids, and subconjunctival hemorrhages. AHC has been reported to be caused by which of the following viruses?

(A) Coronavirus
(B) Reovirus
(C) Rhinovirus
(D) Enterovirus
(E) Respiratory syncytial virus

36. Mumps virus accounts for 10 to 15 percent of all cases of aseptic meningitis in the United States. Infection with mumps virus

(A) is apt to recur periodically in many affected persons
(B) will usually cause mumps orchitis in postpubertal males
(C) is maintained in a large canine reservoir
(D) usually produces severe systemic manifestations
(E) is preventable by immunization

37. The serum of a newborn infant reveals a 1:32 cytomegalovirus (CMV) titer. The child is clinically asymptomatic. Which of the following courses of action would be advisable?

(A) Repeat the CMV titer immediately
(B) Wait 6 months and obtain another titer on the baby
(C) Obtain a CMV titer from all siblings
(D) Obtain an anti-CMV IgM titer from the mother
(E) Obtain an anti-CMV IgM titer from the baby

38. A 3-year-old child presents at the physician's office with symptoms of coryza, conjunctivitis, low-grade fever, and Koplik's spots. The causative agent of this disease belongs to which group of viruses?

(A) Adenovirus
(B) Herpesvirus
(C) Picornavirus
(D) Orthomyxovirus
(E) Paramyxovirus

39. Epidemic pleurodynia and myocarditis of newborn infants are both caused by

(A) group B coxsackievirus
(B) polyomavirus
(C) respiratory syncytial virus
(D) reovirus
(E) cytomegalovirus

Microbiology

40. Rabies virus, which produces one of the most feared of all human diseases, is
    (A) not detected by direct fluorescent antibody microscopy
    (B) not associated with acute ascending paralysis
    (C) the cause of Negri bodies in the cytoplasm of nerve cells
    (D) associated with an incubation period in humans of 7 to 10 days
    (E) not usually fatal in humans

41. Certain viruses have been associated with birth defects. These teratogenic viruses include all the following EXCEPT
    (A) rubella virus
    (B) cytomegalovirus
    (C) coxsackievirus
    (D) herpes simplex virus
    (E) rhinovirus

42. Which virus is the leading cause of the croup syndrome in young children and, when infecting mammalian cells in culture, will hemabsorb red blood cells?
    (A) Group B coxsackievirus
    (B) Rotavirus
    (C) Parainfluenza virus
    (D) Adenovirus
    (E) Rhinovirus

43. After returning from Africa, a woman develops an illness characterized by high fever, black vomitus, proteinuria, and jaundice. She is most likely to have
    (A) dengue
    (B) yellow fever
    (C) kala-azar
    (D) scrub typhus
    (E) equine encephalitis

44. Rotavirus is a double-stranded RNA virus with a double-walled capsid. All the following statements about rotavirus are true EXCEPT
    (A) it is similar to Nebraska calf diarrhea virus
    (B) it is a major cause of neonatal diarrhea
    (C) it is readily cultured from the stool of infected persons
    (D) most newborn infants have maternally acquired antibodies to it
    (E) early breast-feeding helps protect neonates against it

45. All the following viral infections have been associated with paramyxoviruses EXCEPT
    (A) mumps
    (B) measles
    (C) croup
    (D) bronchiolitis
    (E) otitis media

46. The most sensitive method of detecting infection by cytomegalovirus (CMV) in the newborn is

(A) isolation of virus
(B) identification of characteristic cells in gastric secretions
(C) detection of IgM antibody by immunofluorescence
(D) direct detection of antigen by ELISA
(E) detection of complement-fixing antibodies

47. Human papillomavirus is associated with all the following tumors EXCEPT

(A) plantar wart
(B) cervical cancer
(C) condyloma acuminatum
(D) hepatic carcinoma
(E) laryngeal carcinoma

48. Reverse transcriptase is an enzyme unique to the retroviruses. Its functions include all the following activities EXCEPT

(A) RNAase H activity
(B) DNA-dependent DNA polymerase activity
(C) RNA polymerase activity
(D) RNA-dependent DNA polymerase activity
(E) integration activity

49. Which of the following is a DNA virus?

(A) Togavirus
(B) Hepatitis A virus (HAV)
(C) Arenavirus
(D) Parvovirus
(E) Rotavirus

50. St. Louis encephalitis, a viral infection, was first recognized as an entity in 1933. All the following statements regarding this disease are true EXCEPT

(A) it is transmitted to humans by the bite of an infected tick
(B) it is caused by a flavivirus
(C) it is the major arboviral cause of central nervous system infection in the U.S.
(D) it may present as a febrile headache or aseptic meningitis
(E) laboratory diagnosis is usually made by serologic methods

51. Meningitis is characterized by the acute onset of fever and stiff neck. Aseptic meningitis may be caused by a variety of microbial agents. During the initial 24 h of the course of aseptic meningitis, an affected person's cerebrospinal fluid is characterized by

(A) decreased protein content
(B) elevated glucose concentration
(C) lymphocytosis
(D) polymorphonuclear leukocytosis
(E) eosinophilia

52. Rabies is a disease that can have dire consequences in an infected human. Which of the following animals poses the most significant rabies threat to residents of the United States?

(A) Bats
(B) Dogs
(C) Cats
(D) Rats
(E) Mice

53. Infection with hepatitis D virus (HDV; delta agent) can occur simultaneously with infection with hepatitis B virus (HBV) or in a carrier of hepatitis B virus because HDV is a defective virus that requires HBV for its replicative function. What serologic test can be used to determine if a patient with HDV is an HBV carrier?

(A) HBsAg
(B) HBc IgM
(C) HBeAg
(D) HBs IgM
(E) HBs IgG

54. A nurse develops clinical symptoms consistent with hepatitis. She recalls sticking herself with a needle approximately 4 months ago after drawing blood from a patient. Serologic tests for HBsAg, antibodies to HBsAg, and hepatitis A virus (HAV) are all negative; however, she is positive for IgM core antibody. The nurse

(A) does not have hepatitis B
(B) has hepatitis A
(C) is in the late stages of hepatitis B infection
(D) is in the "window" (after the disappearance of HBsAg and before the appearance of anti-HBsAg)
(E) has hepatitis C

55. Human rotavirus is best described by which one of the following statements?

(A) It has a single capsid
(B) It contains segmented plus strand genomes
(C) It is often associated with infantile diarrhea
(D) It uses the host RNA polymerase for mRNA synthesis
(E) Laboratory diagnosis is usually made by culturing the virus from stool

56. All the following statements regarding varicella-zoster virus are true EXCEPT that it

(A) causes shingles
(B) causes chickenpox
(C) is highly contagious
(D) cannot be cultured
(E) belongs to the Herpesviridae family

57. All the following statements about shingles are true EXCEPT

(A) it is caused by the reactivation of latent varicella-zoster virus
(B) the lesions are usually unilateral
(C) exposure to the lesions can produce varicella infection in the nonimmune person
(D) it can be classified as a venereal disease
(E) it can cause chronic disease in the immunocompromised host

58. Adults who have had varicella as children occasionally suffer a recurrent form of the disease, shingles. The agent causing these diseases is a member of which of the following viral families?

(A) Herpesvirus
(B) Poxvirus
(C) Adenovirus
(D) Myxovirus
(E) Paramyxovirus

59. Rhinovirus is primarily transmitted by

(A) droplet aerosolization
(B) sexual activity
(C) fecal-oral route
(D) fomites
(E) vertical transmission from mother to child

60. German measles virus (rubella), a common cause of exanthems in children, is best described by which of the following statements?

(A) Measles (rubeola) and German measles (rubella) are caused by the same virus
(B) Incubation time is approximately 3 to 4 weeks
(C) Vesicular rashes are characteristic
(D) Onset is abrupt with cough, coryza, and fever
(E) Specific antibody in the serum does not prevent disease

61. There is considerable overlap of signs and symptoms seen in congenital and perinatal infections. In a neonate with "classic" symptoms of congenital cytomegalovirus (CMV) infection, all the following tests would be useful in establishing a diagnosis EXCEPT

(A) CMV titer on neonate's serum at birth
(B) CMV titer on mother's serum at birth of infant
(C) CMV titer on neonate's serum at 1 month of age
(D) total IgM on neonate's serum at birth
(E) culture of neonate's urine

62. In most instances, the diagnosis of enteroviral infection is accomplished more quickly and economically by virus isolation than by serologic methods. All the following are acceptable specimens for the isolation of enterovirus EXCEPT

(A) feces
(B) cerebrospinal fluid
(C) throat secretions
(D) urine
(E) skin lesions

63. The presence of Negri inclusion bodies in host cells is characteristic of

(A) mumps
(B) infectious mononucleosis
(C) congenital rubella
(D) aseptic meningitis
(E) rabies

64. Subacute sclerosing panencephalitis (SSPE) is a slow virus infection of humans. As a result of having this disease, a patient may have antibodies capable of reacting with the virus that causes

(A) kuru
(B) scrapie
(C) Creutzfeldt-Jakob disease
(D) measles
(E) progressive multifocal leukoencephalopathy (PML)

65. All the following diseases are characterized as slow virus infections EXCEPT

(A) kuru
(B) subacute sclerosing panencephalitis (SSPE)
(C) Creutzfeldt-Jakob disease
(D) shingles
(E) progressive multifocal leukoencephalopathy (PML)

66. According to recommendations issued by the U.S. Public Health Service, which of the following statements regarding vaccination against smallpox is true?

(A) Pregnant women should be vaccinated in the first trimester
(B) Persons who have eczema should be vaccinated soon after diagnosis
(C) Persons who have immune deficiencies should be vaccinated every 5 years
(D) Persons traveling abroad need not be vaccinated
(E) Children should be vaccinated before they begin school

67. A 4-year-old child presents in the emergency room with a rash and illness characteristic of chickenpox. One member of the medical team attending this patient is a pregnant nurse with no history of chickenpox. The decision is made to administer zoster immune globulin to the nurse, but the suggestion is made that a serum sample be drawn prior to this to determine her true immune status. The test necessary to determine immune status for varicella-zoster virus is

(A) complement fixation
(B) fluorescent antibody to membrane antigen (FAMA)
(C) direct fluorescent antibody
(D) indirect fluorescent antibody
(E) enzyme immunoassay

68. A patient presents with keratoconjunctivitis. The differential diagnosis should include infection with which of the following viruses?

(A) Parvovirus
(B) Adenovirus
(C) Epstein-Barr virus
(D) Respiratory syncytial virus
(E) Varicella-zoster virus

69. Interferon, a protein that inhibits viral replication, is produced by cells in tissue culture when the cells are stimulated with all the following EXCEPT

(A) endotoxin
(B) synthetic double-stranded polynucleotides
(C) viruses
(D) chlamydiae
(E) rickettsiae

70. An 18-year-old female college student has splenomegaly. Serologic examination reveals an elevated white blood cell count (including atypical lymphocytes) and heterophil antibodies. She probably has

(A) mumps
(B) parainfluenza
(C) rubella
(D) infectious mononucleosis
(E) lymphocytic choriomeningitis

71. Since commercially prepared HTLV-I enzyme immunoassay was licensed by the FDA, blood units in the U.S. have been screened for HTLV-I prior to release. This testing would be necessary in all the following components EXCEPT

(A) plasma
(B) whole blood
(C) red blood cells
(D) platelets
(E) white blood cells

72. All the following statements regarding the cytopathic effects of viruses on host cells are true EXCEPT that they are

(A) usually morphologic in nature
(B) often associated with changes in lysosomal membranes
(C) pathognomonic for an infecting virus
(D) usually fatal to the host cell
(E) usually visible with a light microscope

73. A 17-year-old girl presents with cervical lymphadenopathy, fever, and pharyngitis. Infectious mononucleosis is suspected. Clinically useful tests in this diagnosis include all the following EXCEPT

(A) IgM antibody to viral core antigen (VCA)
(B) IgG antibody to VCA
(C) antibody to Epstein-Barr nuclear antigen (EBNA)
(D) culture
(E) heterophil antibody

74. A hospital worker is found to have hepatitis B surface antigen. Subsequent tests reveal the presence of *e* antigen as well. The worker most likely

(A) is infective and has active hepatitis
(B) is infective but does not have active hepatitis
(C) is not infective
(D) is evincing a biologic false positive test for hepatitis
(E) has both hepatitis B and C

75. A hospital patient received 3 units of whole blood and developed severe hepatitis *8 weeks* later. The most probable cause of that hepatitis is

(A) hepatitis A virus
(B) hepatitis B virus
(C) hepatitis D (delta) virus
(D) non-A, non-B hepatitis (hepatitis C) virus
(E) cytomegalovirus

76. HTLV-I and HIV share all the following characteristics EXCEPT

(A) they are nonenveloped viruses
(B) they infect T-cell lymphocytes
(C) they are RNA viruses
(D) they have a reverse transcriptase enzyme
(E) they are retroviruses

77. Influenza is an acute, usually self-limited, febrile illness caused by infection with influenza type A or B virus. A secondary bacterial pneumonia often follows primary viral infection. Commonly involved bacteria include all the following EXCEPT

(A) *Streptococcus pneumoniae*
(B) *Streptococcus pyogenes*
(C) *Staphylococcus aureus*
(D) *Legionella pneumophila*
(E) *Haemophilus influenzae*

78. Viruses known to establish latent infections include all the following EXCEPT

(A) adenovirus
(B) herpes simplex virus (HSV)
(C) cytomegalovirus (CMV)
(D) varicella-zoster virus (VZV)
(E) coxsackievirus group B

79. A regimen that includes appropriately administered gamma globulin may be indicated in all the following EXCEPT

(A) hepatitis A
(B) hepatitis B
(C) rabies
(D) poliomyelitis
(E) infectious mononucleosis

80. Alphavirus causes which one of the following viral diseases?

(A) Marburg virus disease
(B) St. Louis encephalitis
(C) Western equine encephalitis
(D) Dengue fever
(E) Yellow fever

81. Diseases in which atypical lymphocytosis may be found include all the following EXCEPT

(A) encephalitis caused by herpes simplex virus (HSV)
(B) mononucleosis induced by Epstein-Barr virus
(C) cytomegalovirus infection
(D) acute viral hepatitis
(E) toxoplasmosis

82. All the following statements concerning the Western blotting test for the detection of antibody to HIV are true EXCEPT

(A) viral protein is blotted onto nitrocellulose paper
(B) indeterminate results may be obtained
(C) antibody to individual viral proteins may be visualized
(D) it is a good screening test for HIV infection
(E) it is more specific than the ELISA assay for detecting antibody for HIV

83. Several antiviral compounds have been developed during the last decade. One such compound is Ribavirin, a synthetic nucleoside structurally related to guanosine. Ribavirin therapy has been successfully used against

(A) respiratory syncytial virus
(B) herpes simplex virus
(C) rhinovirus
(D) group A coxsackievirus
(E) parvovirus

84. A defective virus is one that lacks one or more functional genes required for viral replication. Which of the following viruses is defective?

(A) Herpes simplex type 2
(B) Cytomegalovirus
(C) Hepatitis A virus
(D) Hepatitis D virus
(E) Epstein-Barr virus

85. In its early stages, dengue is characterized by all the following EXCEPT

(A) arthralgia
(B) maculopapular rash
(C) lymphadenopathy
(D) edema
(E) fever

86. Viruses that carry enzymes for mRNA synthesis as part of their virion and introduce them into host cells upon infection include

(A) reoviruses
(B) retroviruses
(C) rotaviruses
(D) togaviruses
(E) herpesviruses

87. Hepatitis C (HCV) is usually a clinically mild disease, with only minimal elevation of liver enzymes. Hospitalization is unusual. All the following statements regarding HCV are true EXCEPT

(A) 30 to 50 percent of cases progress to chronic liver disease
(B) it often occurs in posttransfusion patients
(C) many HCV infections occur in IV drug abusers
(D) it is a DNA virus
(E) blood products can now be tested for antibody to HCV

Microbiology

DIRECTIONS: Each group of questions below consists of lettered headings followed by a set of numbered items. For each numbered item select the **one** lettered heading with which it is **most** closely associated. Each lettered heading may be used **once, more than once, or not at all.**

Questions 88–91

The antigens and antibody below are associated with hepatitis. For each, choose the description with which it is most likely to be associated.

(A) Is usually the first viral marker detected in blood after HBV infection
(B) May be the only detectable serologic marker during the early convalescent phase of an HBV infection ("window phase")
(C) Appears in the blood soon after infection, rises to very high concentrations, and falls rapidly with the onset of hepatic disease
(D) Found within the nuclei of infected hepatocytes and not generally in the peripheral circulation
(E) Closely associated with hepatitis B infectivity and DNA polymerase activity

88. HBeAg

89. HBsAg

90. HBcAg

91. Anti-HBc

Questions 92–96

For each of the viral diseases below, choose the source from which the vaccine for that disease is obtained.

(A) Calf or sheep lymph
(B) Duck embryo
(C) Chick embryo cell culture
(D) Chick embryo tissue culture
(E) Monkey kidney tissue culture

92. Eastern equine encephalitis

93. Mumps

94. Measles

95. Rabies

96. Smallpox

Virology

Questions 97–101

For each of the preparations below, choose the virus or disease to which it is most closely related.

(A) Hepatitis A
(B) Influenza A
(C) Measles
(D) Herpes simplex
(E) Hepatitis B

97. Acyclovir

98. Pooled serum immune globulin

99. Killed virus vaccine

100. Live virus vaccine

101. Recombinant viral vaccine

Questions 102–106

For each disease listed below, select the virus most likely to be the causative agent.

(A) Cytomegalovirus
(B) Rotavirus
(C) Varicella-zoster virus
(D) Adenovirus
(E) Papillomavirus

102. Chickenpox

103. Acute respiratory disease (ARD)

104. Human warts

105. Infantile gastroenteritis

106. Heterophil-negative infectious mononucleosis

Microbiology

DIRECTIONS: Each group of questions below consists of four lettered headings followed by a set of numbered items. For each numbered item select

A   if the item is associated with   (A) **only**
B   if the item is associated with   (B) **only**
C   if the item is associated with   **both** (A) and (B)
D   if the item is associated with   **neither** (A) nor (B)

Each lettered heading may be used **once, more than once, or not at all.**

Questions 107–111

(A) Hepatitis A virus
(B) Hepatitis B virus
(C) Both
(D) Neither

107. IgM antibody to the viral particle is important in laboratory diagnosis

108. Passive immunity is conferred with immune serum globulin (ISG)

109. Vaccination with viral surface antigen usually provides immunity

110. DNA viral genome is contained

111. RNA viral genome is contained

Questions 112–115

(A) Human papillomaviruses
(B) Polyomaviruses
(C) Both
(D) Neither

112. Causative agents of a variety of cutaneous warts (plantar, common, and flat) and associated with cervical neoplasia

113. Nonenveloped, icosahedral DNA viruses

114. Progressive multifocal leukoencephalopathy (PML), a disease causing demyelination in the central nervous system

115. Cryotherapy as the most popular therapy

Questions 116–120

(A) Orthomyxoviruses
(B) Paramyxoviruses
(C) Both
(D) Neither

116. Single-stranded RNA virus

117. Influenza A virus, influenza B virus, and mumps

118. Viral-coded RNA polymerase incorporated into virus

119. Measles virus, respiratory syncytial virus (RSV), and parainfluenza virus

120. Replication exclusively in the nucleus of the host cell

Virology
Answers

1. **The answer is B.** *(Jawetz, 19/e. p 578.)* If the patient was infected 2 weeks prior, there is little likelihood of a detectable antibody response. Therefore, either EIA or Western blotting for antibody could be negative in spite of infection. All three of the other tests determine the presence of virus and might be appropriate. However, culture generally costs around $300 and is only 80 percent sensitive. Polymerase chain reaction is a very sensitive test if done correctly, but it costs about $150. The antigen test is the best, most cost-efficient test available.

2. **The answer is D.** *(Jawetz, 19/e. pp 539–544.)* Rabies virus produces an acute neurologic infection carrying a fatality rate of almost 100 percent. Animals can transmit the virus for only a few days (e.g., 5 days for dogs) before their disease becomes apparent; the bat is a notable exception to this rule. Rabies virus, which is found in just one antigenic type, can be isolated from the saliva of infected persons.

3. **The answer is A.** *(Davis, 4/e. pp 1071–1073.)* Although infection with cytomegalovirus (CMV) is common, it only rarely causes clinically apparent disease. Lesions characteristic of infection with CMV are found in up to 10 percent of stillborn babies; however, CMV, which can be transmitted transplacentally, usually is not the cause of death. Children and adults with immunosuppressive problems are susceptible to active disease. In severely immunodeficient patients such as those with AIDS, CMV ocular disease may occur. The patient suffers blurring of vision or vision loss and ophthalmic examination reveals large yellowish-white areas with flame-shaped hemorrhages. Ganciclovir is now licensed for treatment of CMV retinitis in AIDS patients.

4. **The answer is E.** *(Jawetz, 19/e. pp 523–525.)* Respiratory syncytial virus (RSV), the major cause of lower respiratory tract illness in young children, can be diagnostically confirmed by isolation of the virus from respiratory secretions. Because the virus exhibits extreme lability, the highest rate of viral recovery is obtained by the immediate inoculation of a nasal wash into Hep-2 cells. An immunofluorescent technique for rapid diagnosis can be used directly on clinical specimens.

Virology Answers 21

5. **The answer is C.** *(Jawetz, 19/e. pp 434–438.)* Cytomegalovirus (CMV) can cause infectious mononucleosis–like disease. However, heterophil antibodies are usually absent. This disease can either occur spontaneously or after blood transfusion. CMV has been isolated from peripheral blood leukocytes of patients with this disease.

6. **The answer is A.** *(Davis, 4/e. pp 992–994.)* Influenza A viruses have undergone three major antigenic shifts since the 1933 pandemic, and each shift has been accompanied by a high degree of susceptibility in a large part of the population. This antigenic shift affects both the hemagglutinin and neuraminidase antigens and creates serious problems in developing effective vaccines. Minor antigenic changes (antigenic drift), which occur every 2 to 3 years, have had less impact because of cross-reactivity of the neutralizing antibodies. Variations in influenza B viruses have been less marked and less frequent and have all been subject to immune cross-reactivity.

7. **The answer is C.** *(Davis, 4/e. pp 927–928.)* Parvovirus B19 is the causative agent of erythema infectiosum (fifth disease). It is associated with transient aplastic crisis in persons with hereditary hemolytic anemia. In adults it is also associated with polyarthralgia.

8. **The answer is C.** *(Davis, 4/e. p 935.)* The initial infection by herpes simplex virus is often inapparent and occurs through a break in the skin or mucous membranes, such as in the eye, throat, or genitals. Latent infection often persists at the initial site despite high antibody titers. Recurrent disease can be triggered by temperature change, emotional distress, and hormonal factors. Type 1 herpes simplex virus is usually, but not exclusively, associated with ocular and oral lesions; type 2 is usually, but not exclusively, associated with genital and anal lesions. Type 2 infection is more common. In addition to mucocutaneous infections, the CNS and occasionally visceral organs can be involved.

9. **The answer is B.** *(Jawetz, 19/e. pp 483–486.)* Rotaviruses were initially identified by direct electron microscopy (EM) of duodenal mucosa of infants with gastroenteritis. Subsequent studies in several countries have shown them to be the cause of 30 to 40 percent of acute diarrhea in infants. They are non-lipid-containing RNA viruses with a double-shelled capsid. Although the virus has been serially propagated in human fetal intestinal organ cultures, cytopathic changes are minimal or absent; multiplication is detected by immunofluorescence. Numerous methods for rotavirus antigen detection, including radioimmunoassay, counterimmunoelectrophoresis, and enzyme-linked immunosorbent assay, have been developed and found to be about as effective as EM.

10. The answer is C. *(Jawetz, 19/e. pp 418–427.)* Herpes simplex virus (HSV) is an extremely common human pathogen. It causes a variety of infections ranging from asymptomatic infection to fulminant disseminated disease resulting in death. All herpes viruses contain a double-stranded DNA genome. The salient feature of HSV is its ability to exist in the latent stage following primary infection. Treatment with acyclovir has been most beneficial during primary infection; however, treatment of reactivated virus infection has resulted in shorter duration of individual episodes. There is no vaccine for HSV.

11. The answer is C. *(Davis, 4/e. pp 1145–1147.)* *Tat* encodes a transactivator that increases production of active messenger RNA and may act as an antiterminator, while *rev* affects posttranscriptional activities. The products of the two genes greatly increase viral replication. The *gag* and *pol* genes both produce structural proteins. The *nef* protein down-regulates the transcription of the HIV genome.

12. The answer is A. *(Balows, Clinical Microbiology, 5/e. pp 985–996.)* Marburg and Ebola belong to the Filoviridae family. Infections with these two viruses carry a high mortality and have no effective treatment. If either of these agents is suspected, barrier protection and isolation of the patient are necessities. Culture of the virus may be attempted only at a biosafety level 4 laboratory, which can provide maximum containment. Recently, nonhuman primates were infected with Ebola virus and became ill during quarantine.

13. The answer is C. *(Davis, 4/e. pp 1091–1092.)* Molluscum contagiosum virus causes a rare skin disease primarily affecting children and young adults. The virus stimulates cell division in uninfected neighboring cells, even though DNA synthesis ceases in infected cells. It has been transmitted experimentally to humans and has been grown in cultures of HeLa cells and human amnion. The chronic lesion is restricted to the epithelium of the skin of the back, face, arms, legs, and genitals. Electron microscopy shows virions that are identical to those of other poxviruses.

14. The answer is B. *(Davis, 4/e. pp 1149–1150.)* Homosexual acts between males constitute one of the highest risk activities in the U.S. However, transmission is not restricted to male homosexuals; HIV is also transmitted heterosexually. Transmission by infected blood is recognized as another major cause of infection among IV drug abusers who share needles. Transplacental and prenatal transmission is also documented with infants born to infected mothers. While homosexual transmission between females has been reported, sexually active homosexual females are not considered to be at high risk at this time.

Virology Answers

15. The answer is A. *(Balows, Laboratory Diagnosis, pp 384–392.)* Rotavirus, a segmented RNA virus, is a common nosocomial pathogen especially in children's hospitals and nurseries. Culture requires trypsin activation of cell lines and omission of serum that contains inhibitors of trypsin. Even with special treatment, no cytopathic effect (CPE) is observed and immunofluorescent staining is necessary. Because culture is difficult and antigen excretion is high, the method of choice for diagnosis of rotavirus infection is detection of antigen by either latex agglutination or enzyme immunoassay.

16. The answer is B. *(MMWR 37:737–738, 1988.)* Adult T-cell leukemia/lymphoma (ATL) is a malignancy of mature T-lymphocytes. It is recognized in Japan, the Caribbean, and Africa. Of 74 cases reported in the U.S. between 1980 and 1987, half involved persons of Japanese or Caribbean ancestry. ATL is also associated with a degenerative neurologic disease known in the Caribbean as *tropical spastic paraparesis* and in Japan as *HTLV-I–associated myelopathy*.

17. The answer is C. *(Balows, Clinical Microbiology, 5/e. pp 848–850.)* All of Koch's postulates have been verified for the relationship between infectious mononucleosis and Epstein-Barr virus, a herpesvirus. However, the relationship between this virus and Burkitt's lymphoma, sarcoid, and systemic lupus erythematosus (SLE) is less clear. Infectious mononucleosis is most common in young adults (14 to 18 years of age) and is very rare in young children. There is no specific treatment. Heterophil antibody titer is helpful in diagnosis, but is not expressed as a function of clinical recovery.

18. The answer is A. *(Jawetz, 19/e. pp 536–538.)* Coronaviruses are enveloped viruses containing a genome of single-stranded RNA. Their nucleocapsids develop in the cytoplasm and mature by budding into cytoplasmic vesicles. These viruses are commonly isolated from patients with upper respiratory tract illnesses during the winter months. They have also been implicated as a cause of gastroenteritis in children.

19. The answer is B. *(Jawetz, 19/e. pp 531–535.)* Rubella virus does not produce cytopathic effects (CPEs) in tissue-culture cells. Moreover, rubella-infected cells challenged with a picornavirus are resistant to subsequent infection and thus would not exhibit CPEs. Mouse kidney cells infected only with picornavirus would show CPEs.

20. The answer is D. *(Balows, Laboratory Diagnosis, p 8.)* The maxim "as temperature increases, viral titers decrease" can be used as a general rule. However, $-20°C$ is not a good temperature for transport of viral specimens because slow freezing allows formation of ice crystals, which are deleterious to the virus. Long-term storage is possible at −70°C, however, because rapid freezing occurs.

21. The answer is A. *(Jawetz, 19/e. pp 488–502.)* Arboviruses (*a*rthropod-*bo*rne viruses) may or may not be surrounded by a lipid envelope, although most are inactivated by lipid solvents such as ether, and may contain either double-stranded or single-stranded RNA. Physicochemical studies have demonstrated a great heterogeneity among these viruses. Arboviruses cause disease in vertebrates; in humans, encephalitis is a frequent arbovirus illness. Most human infections with arbovirus, however, are asymptomatic.

22. The answer is E. *(Balows, Laboratory Diagnosis, p 529.)* SSPE is a late and rare manifestation of measles. It is a progressive encephalitis involving both white and gray matter. Demyelination is seen only at an advanced stage of the disease in a few cases. In 1985, viral RNA was demonstrated in brain cells from a patient with SSPE by the use of in situ hybridization.

23. The answer is D. *(Jawetz, 19/e. pp 401–403.)* Interferon is a protein produced by cells in response to a viral infection or certain other agents. Entering uninfected cells, interferon causes production of a second protein that alters protein synthesis. As a result of inhibition of either translation or transcription, new viruses are not assembled following infection of interferon-protected cells.

24. The answer is D. *(Balows, Laboratory Diagnosis, p 692.)* A variety of RNA-containing viruses belong to the family Picornaviridae. The organisms are transmitted by the fecal-oral route and can cause a variety of illnesses, such as paralytic polio (poliovirus); aseptic meningitis (enterovirus); hand, foot and mouth disease (coxsackie A16); and myocarditis (coxsackie B). Infections with viruses belonging to Picornaviridae are quite common, but severe clinical manifestations as mentioned above are not unusual.

25. The answer is B. *(Davis, 4/e. pp 1162–1165.)* Coronaviruses, discovered in 1965, are thought to be a major agent of the common cold, especially in older children and adults. The virion is known to contain RNA, but other elements of its structure are unclear. At 34°C viral multiplication is profuse; however, infectivity is greatly reduced at higher temperatures or following extended incubation.

26. The answer is A. *(Jawetz, 19/e. p 512.)* Neither the pathogens nor the pathophysiology of Reye's syndrome has been elucidated. Mild viral infections usually precede the onset of Reye's syndrome. The majority of cases occur during outbreaks of influenza B; however, influenza A viruses have also been isolated from patients with Reye's syndrome.

Virology Answers

27. The answer is A. *(Jawetz, 19/e. pp 435–436.)* Fifty to eighty percent of patients who have infectious mononucleosis develop an increased titer of heterophil antibodies that can be demonstrated by agglutination of sheep erythrocytes. Affected persons have lymphadenopathy, and the presence of atypical lymphocytes in the peripheral smear is characteristic. The Epstein-Barr virus is thought to be the etiologic agent.

28. The answer is D. *(Balows, Laboratory Diagnosis, pp 438–446.)* While the highest risk of fetal infection with rubella occurs during the first trimester of pregnancy (90 to 100 percent chance of infection), there is still a 10 to 20 percent risk as late as the 16th week of gestation. Between weeks 17 and 20, occasional hearing loss and retinopathy occur. Vaccination is contraindicated and rubella immune serum does not exist. Serologic testing allows documentation of infection and calculation of risk for the fetus.

29. The answer is B. *(Jawetz, 19/e. pp 517–535.)* Hemadsorption activity of inoculated cultures suspected of containing paramyxovirus (e.g., mumps virus) should be tested at 4°C. At this temperature the enzyme neuraminidase is unable to exert its inhibitory effect on hemagglutination. At 35°C, on the other hand, neuraminidase is active and destroys hemagglutinin receptors.

30. The answer is D. *(Jawetz, 19/e. pp 452, 456.)* The delta agent was first described in 1977 and has recently been shown to be an incomplete RNA virus that requires HBsAg for replication. It is found most often in persons who have multiple parenteral exposures, e.g., intravenous (IV) drug abusers, hemophiliacs, and multiply transfused patients.

31. The answer is C. *(Jawetz, 19/e. pp 525–527.)* Both mumps and measles are diseases caused by paramyxoviruses, which are enveloped and contain single-stranded RNA. Rabies is caused by a rhabdovirus, variola (smallpox) by a poxvirus, and varicella (chickenpox) by a herpesvirus. The causative agent of hepatitis is as yet unclassified.

32. The answer is D. *(Davis, 4/e. p 881.)* As an intravenous agent, ribavirin is effective against Lassa fever in the first week of onset of the disease. It may also be administered as an aerosol that is quite useful in infants with RSV. Unlike amantadine, which is efficacious only with influenza A, ribavirin has activity against both influenza A and B if administered by aerosol in the first 24 h of onset.

33. The answer is D. *(Jawetz, 19/e. pp 447–479.)* Echoviruses were discovered accidentally during studies on poliomyelitis. They were named "*e*nteric *c*ytopathogenic *h*uman *o*rphan (ECHO) viruses" because, at the time, they had not been linked to human disease and thus were considered "orphans."

Echoviruses now are known to infect the intestinal tract of humans; they also can cause aseptic meningitis, febrile illnesses, and the common cold. Echoviruses range in size from 24 to 30 nm in diameter and contain a core of RNA.

34. The answer is D. *(Jawetz, 19/e. pp 422–430.)* Acute herpetic gingivostomatitis is the most common clinical disease caused by primary infection with type 1 herpes simplex virus. The disorder is characterized by fever, local lymphadenopathy, and extensive lesions of the mucous membranes of the oral cavity. This stomatitis occurs most frequently in young children. The lesions may recur, repeatedly and at various intervals, in the same location.

35. The answer is D. *(Jawetz, 19/e. pp 471–479.)* Enterovirus and coxsackievirus A can be recovered from conjunctival scrapings of patients with acute hemorrhagic conjunctivitis (AHC) during the first 3 days of illness. Isolation rates are somewhat higher for enterovirus than coxsackievirus. Less than 5 percent of throat swab or fecal specimens have been positive for either virus.

36. The answer is E. *(Jawetz, 19/e. pp 525–527.)* Much of the public's understanding of mumps is based on suppositions that are without any scientific basis. For example, natural mumps infection confers immunity after a single infection, even if the infection was a unilateral, not bilateral, parotitis. Also, sterility from mumps orchitis is not assured; only 20 percent of males older than 13 years of age develop orchitis. The majority of patients with mumps do not develop systemic manifestations. In fact, some do not develop parotitis. Lastly, the virus is maintained exclusively in human populations; canine reservoirs are not known. The mumps vaccine is a live attenuated virus vaccine derived from chick-embryo tissue culture.

37. The answer is E. *(Jawetz, 19/e. pp 430–434.)* Clinical manifestations of cytomegalovirus (CMV) infection may not be readily apparent at birth. Thus, in a newborn infant with a 1:32 titer of CMV, it is necessary to determine whether the antibodies were passed transplacentally from the mother (these antibodies would be IgG) or produced by the fetus in response to an in utero infection (IgM). A newborn infant who is infected excretes large numbers of virus particles in the urine and, therefore, places other neonates at risk for contracting CMV disease.

38. The answer is E. *(Davis, 4/e. p 1018.)* Koplik's spots are pathognomonic for measles. The measles virus is a paramyxovirus. In industrialized countries, vaccination has reduced the importance of this childhood infection (although U.S. incidence increased in 1989 and 1990). In developing countries, however, measles is a major killer of young children. In America, most states

Virology Answers 27

now require proof of immunity before school enrollment, and this has reduced the incidence of disease.

39. The answer is A. *(Jawetz, 19/e. pp 475–477.)* The coxsackieviruses (groups A and B) produce a variety of illnesses, including aseptic meningitis, acute upper respiratory disease, and a paralytic disease simulating poliomyelitis. Twice the normal incidence of congenital heart lesions is found in infants whose mothers had coxsackievirus infections during the first trimester of pregnancy. Epidemic pleurodynia (Bornholm disease, epidemic myalgia) is a coxsackieviral disease causing paroxysmal chest pain and increasing fever and debility.

40. The answer is C. *(Balows, Laboratory Diagnosis, pp 575–579. Jawetz, 19/e. pp 457–461.)* Rabies, an extremely virulent disease in humans, is an ether-sensitive RNA virus (a rhabdovirus) that forms Negri bodies in the cytoplasm of infected nerve cells. Chiropteran, or bat, rabies, which may differ from canine strains, can produce an ascending myelitis and spreading paralysis that resembles acute ascending paralysis (Landry's paralysis) or the Guillain-Barré syndrome. It is interesting to note that this type of paralysis has been reported to follow antirabies inoculation. The incubation period of human rabies can be for extended periods of time. The introduction of fluorescent antibody staining of rabies antigen enabled rapid diagnosis of rabid animals. Specimens from animals should routinely include the brainstem, cerebellum, and hippocampus. Many areas of the U.S. are now seeing cat and dog rabies. The first case of cat and dog rabies in Connecticut in 30 years was recently reported (1992).

41. The answer is E. *(Davis, 4/e. pp 903–904.)* Several viruses are known to be teratogenic, i.e., to produce congenital malformations in humans. Rubella infection during pregnancy is associated with fetal anomalies, and rubella remains the major viral cause of fetal death. Cytomegalovirus may cause microcephaly, herpesvirus may lead to CNS disease, and coxsackievirus may produce cardiac lesions.

42. The answer is C. *(Jawetz, 19/e. pp 521–523.)* Parainfluenza viruses are important causes of respiratory diseases in infants and young children. The spectrum of disease caused by these viruses ranges from a mild febrile cold to croup, bronchiolitis, and pneumonia. Parainfluenza viruses contain RNA in a nucleocapsid encased within an envelope derived from the host cell membrane. Infected mammalian cell culture will hemabsorb red blood cells owing to viral hemagglutinin on the surface of the cell.

43. The answer is B. *(Jawetz, 19/e. pp 497–499.)* Yellow fever is an arbovirus infection transmitted by the bite of the *Aedes* mosquito. A disease of South

America and Africa, yellow fever is characterized by a high fever, jaundice, proteinuria, and black vomitus. Dengue, also caused by an *Aedes*-borne arbovirus, is usually a more benign infection than yellow fever and is characterized by fever, muscle and joint pain, and lymphadenopathy. Equine encephalitis is caused by yet another arbovirus, kala-azar by a parasite, and scrub typhus by rickettsiae.

44. The answer is C. *(Volk, 3/e. pp 708–709.)* Rotavirus is a viral entity that is similar to Nebraska calf diarrhea virus and is thought to be a major cause of acute diarrhea in newborn infants. Three-quarters of all adults have antibodies against rotavirus; passive transfer of these antibodies to the baby, especially through the colostrum, seems to be protective. Although vaccination would be expected to be of little use to the neonate, it might effectively immunize pregnant mothers.

45. The answer is E. *(Balows, Laboratory Diagnosis, pp 484–545.)* Both mumps and measles are well-recognized paramyxovirus infections. This group also includes parainfluenza virus, which causes laryngotracheobronchitis (croup) in children, and respiratory syncytial virus, which can cause bronchiolitis in infants. Paramyxoviruses have glycoprotein spikes that extend their lipid membrane and are responsible for hemagglutination activities.

46. The answer is A. *(Davis, 4/e. p 941.)* While all the listed methods can be used to detect infection by CMV, none is as sensitive as viral cultivation. Identification of characteristic cells offers an inexpensive method of identifying infection by CMV, and detection of antigens offers a more rapid method than cultivation. Detection of IgM by immunofluorescence is valuable in determining neonatal infection as it will identify the baby's antibody response rather than maternal IgG.

47. The answer is D. *(Balows, Laboratory Diagnosis, pp 301–305.)* Papillomavirus infects the skin or mucosa and causes benign tumors. These tumors may undergo malignant conversion and become squamous cell carcinomas. Classification of the human papillomavirus is done by DNA hybridization, and to date 46 types have been recognized. Some types, such as 16 and 18, are more frequently associated with carcinoma, while others, such as 6 and 11, are associated with benign tumors or warts.

48. The answer is C. *(Davis, 4/e. p 1128.)* The replication of a retroviral genome is dependent on the reverse transcriptase enzyme, which performs a variety of functions. It builds a complementary strand of DNA for the viral RNA template; it builds a second DNA strand complementary to the previous DNA; it degrades the original RNA, leaving a DNA-DNA duplex; and, fi-

Virology Answers

nally, it is responsible for integrating the new viral DNA hybrid into the host genome.

49. The answer is D. *(Balows, Laboratory Diagnosis, p 160.)* Parvovirus is a DNA virus with no envelope. Togavirus is an enveloped, single-stranded RNA virus, while arenavirus and rotavirus also have an RNA genome. Most DNA viruses are double-stranded, while most RNA viruses are single-stranded. The exceptions are parvovirus (single-stranded DNA) and reovirus and orbivirus (double-stranded RNA). Hepatitis A virus is a single-stranded RNA virus.

50. The answer is A. *(Jawetz, 19/e. pp 493–494.)* St. Louis encephalitis virus has structural and biologic characteristics in common with other flaviviruses. It is the most important arboviral disease in North America. St. Louis encephalitis virus was first isolated from mosquitoes in California. Patients who contract the disease usually present with one of three clinical manifestations: febrile headache, aseptic meningitis, or clinical encephalitis.

51. The answer is D. *(Jawetz, 19/e. p 544.)* Aseptic meningitis is characterized by a pleocytosis of mononuclear cells in the cerebrospinal fluid; polymorphonuclear cells predominate during the first 24 h, but a shift to lymphocytes occurs thereafter. The cerebrospinal fluid of affected persons is free of culturable bacteria and contains normal glucose and slightly elevated protein levels. Peripheral white blood cell counts usually are normal. Although viruses are the most common cause of aseptic meningitis, spirochetes, chlamydiae, and other microorganisms also can produce the disease.

52. The answer is A. *(Jawetz, 19/e. pp 539–544.)* Since 1960, the yearly incidence of human rabies in the United States has ranged from one to three cases. Wild animals—particularly bats, skunks, and foxes—are the chief offenders, accounting for 70 percent of the 3123 cases of animal rabies reported in the United States in 1974. Worldwide, 1000 fatal human cases or more are reported annually. Dogs and cats are the most common sources of human exposure to rabies in most countries of the world other than the United States. Rabies has also been reported in at least two corneal transplant patients who apparently contracted the disease from infected eye donors.

53. The answer is B. *(Balows, Laboratory Diagnosis, p 779.)* In a chronic HBV carrier, there would be no HB core IgM antibody, whereas it would be present in a new HBV infection. The HBe antigen could be present in either an HBV carrier or in acute infection. HBsAg would be present in either a new infection or in the carrier state, while HBsAb would not be present in either case.

54. The answer is D. *(Jawetz, 19/e. pp 461–462.)* In a small number of patients with acute hepatitis B infection, HBsAg can never be detected. In others, HBsAg becomes negative before the onset of the disease or before the end of the clinical illness. In such patients with acute hepatitis, hepatitis B virus infection may only be established by the presence of anti-hepatitis B core IgM (anti-HBc IgM), a rising titer of anti-HBc, or the subsequent appearance of anti-HBsAg.

55. The answer is C. *(Jawetz, 19/e. pp 483–486.)* Rotavirus, a member of the family Reoviridae, has a double capsid and contains no lipid. Rotavirus gastroenteritis is mainly a disease of infants. It usually occurs between 4 months and 3 years of age. Viral particles can be detected in the stool of infected infants by electron microscopy.

56. The answer is D. *(Mandell, 3/e. pp 1139–1144.)* Initial infection with varicella-zoster virus results in chickenpox. The virus then enters a latent phase and may later manifest itself as herpes zoster, or shingles. The virus is highly contagious, especially when manifested as chickenpox. Vesicular fluid can be cultured during the first 3 days of rash by inoculation onto cell lines, such as human foreskin.

57. The answer is D. *(Mandell, 3/e. pp 1156–1158.)* Zoster (shingles) is most common in adults and is characterized by a reactivation of the virus in the posterior nerve roots and ganglia. This is accompanied by vesicles on the skin directly above the affected sensory nerves. The erupted vesicles are usually unilateral; the trunk, head, and neck are most commonly involved. Zoster in children or adults can be the source of varicella in children. Varicella-zoster virus is morphologically identical to herpes simplex virus.

58. The answer is A. *(Jawetz, 19/e. pp 422–439.)* Varicella-zoster virus, a member of the herpesvirus group, causes a usually mild, self-limited illness in children. Recurrent disease in adults who possess circulating antibody against varicella-zoster virus may be more severe and cause an inflammatory reaction in the sensory ganglia of spinal or cranial nerves. This disease, shingles, appears to result from the reactivation (by trauma or other stimuli) of latent varicella-zoster virus.

59. The answer is D. *(Balows, Laboratory Diagnosis, pp 729–730.)* Rhinovirus is a major cause of the common cold. The primary mode of transmission is the contact of contaminated hands, fingers, or fomites with the conjunctiva or nasal epithelium. While several studies have shown no evidence of aerosol transmission, a study by Dick and associates in 1986 did show aerosol transmission can occur. This is not, however, the main mode of transmission.

60. The answer is D. *(Jawetz, 19/e. pp 527–535.)* Measles (rubeola) is an acute, highly infectious disease characterized by a maculopapular rash. German measles (rubella) is an acute, febrile illness characterized by a rash as well as suboccipital lymphadenopathy. Incubation time is 9 full days after exposure. Onset is abrupt and symptoms mostly catarrhal. Koplik's spots, pale, bluish-white spots in red areolas, can frequently be observed on the mucous membranes of the mouth and are pathognomonic for measles.

61. The answer is D. *(Mandell, 3/e. pp 1159–1168.)* Presently, cytomegalovirus (CMV) is the most common cause of congenital and perinatal viral infections. Culture of the virus is a sensitive diagnostic technique; in the case of a neonate with "classic" symptoms, serum samples from the mother and neonate are obtained at birth. The antibody titer in the infant's serum should be higher than the mother's titer, but they may be similar. For this reason, another sample from the infant at 1 month of age is tested simultaneously with the initial sample. The results should indicate a rise in titer. Measurement of total IgM in the infant's sera at birth is nonspecific and may show false negative and false positive reactions.

62. The answer is D. *(Jawetz, 19/e. pp 471–479.)* For the diagnosis of enteroviral infection, the most useful information is obtained when cultures are obtained from multiple sites, which should include throat secretions and feces, because the virus is often not present simultaneously in all specimens. The most definitive evidence is provided by virus isolation from sources such as cerebrospinal fluid, pericardial fluid, or skin lesions, as dictated by the clinical syndrome. Serologic testing has limited usefulness in the diagnosis of nonpolio enteroviral infections because of the large number of virus serotypes.

63. The answer is E. *(Jawetz, 19/e. pp 539–544.)* The definitive diagnosis of rabies in humans is based on the finding of Negri bodies, which are cytoplasmic inclusions in the nerve cells of the spinal cord and brain, especially in the hippocampus. Negri bodies are eosinophilic and generally spherical in shape; several may appear in a given cell. Negri bodies, although pathognomonic for rabies, are not found in all cases of the disease.

64. The answer is D. *(Jawetz, 19/e. p 546.)* Slow viruses produce progressive neurologic disease and may have incubation periods of up to 5 years before their clinical manifestations become apparent. Progressive multifocal leuko-

Microbiology

encephalopathy (PML), subacute sclerosing panencephalitis (SSPE), kuru, and Creutzfeldt-Jakob disease are human diseases caused by slow viruses; other chronic diseases undoubtedly will someday prove to be of similar origin. SSPE apparently is caused by a virus closely resembling the virus that causes measles, and affected persons might possess antibodies against the measles virus. Scrapie is one of several slow virus diseases of animals.

65. The answer is D. *(Mandell, 3/e. pp 769–775, 1200–1201.)* Slow virus refers to those viruses requiring prolonged periods of infection before chronic disease appears. Kuru, a degenerative brain disease, was described among the Fore tribe of New Guinea; patients were infected as children through ritualistic cannibalism, while the manifestation of the disease occurred years later. SSPE is an unusual and late manifestation of measles. Creutzfeldt-Jakob disease is a slow virus infection that causes presenile dementia. Shingles is merely a manifestation of a latent infection with varicella-zoster virus. PML is a rare, subacute, and progressive demyelinating disease of the central nervous system.

66. The answer is D. *(Jawetz, 19/e. pp 443–447.)* Routine vaccination of infants and children for smallpox has been discontinued in the United States, both because the risk of contracting the disease is so low and because the complications of smallpox vaccination, including generalized vaccinia eruption, postvaccine encephalitis, and fetal vaccinia, are significant. Owing to the extremely effective eradication of smallpox worldwide by the World Health Organization, U.S. citizens traveling abroad no longer require vaccination. Pregnancy, immune deficiencies, and eczema and other chronic dermatitides are contraindications to smallpox vaccination.

67. The answer is B. *(Balows, Laboratory Diagnosis, pp 270–272.)* Complement fixation and the indirect fluorescent antibody test are both useful in diagnosing acute infections with varicella-zoster virus (VZV); however, they do not have the sensitivity to determine immune status. The direct fluorescent antibody test is a test for antigen, not antibody. Enzyme immunoassay is not a procedure available for VZV. FAMA is usually the best assay for determination of immune status in VZV. Also, neutralization and anticomplement immunofluorescent tests may be used.

68. The answer is B. *(Balows, Laboratory Diagnosis, pp 211, 290.)* Adenovirus type 8 is associated with epidemic keratoconjunctivitis, while adenovirus types 3 and 4 are often associated with "swimming pool conjunctivitis." There are also reports of nosocomial conjunctivitis with adenovirus. Herpes simplex virus can infect the conjunctiva and is among the most common causes of blindness in North America and Europe.

Virology Answers

69. The answer is D. *(Jawetz, 19/e. pp 401–403.)* Interferon is a protein that alters cell metabolism to inhibit viral replication. It induces the formation of a second protein that interferes with the translation of viral messenger RNA. Production of interferon has been demonstrated when cells in tissue culture are challenged with viruses, rickettsiae, endotoxin, or synthetic double-stranded polynucleotides. Interferon confers species-specific, not virus-specific, protection for cells.

70. The answer is D. *(Jawetz, 19/e. pp 434–438.)* Infectious mononucleosis characteristically is accompanied by splenomegaly, the appearance of unique sheep-cell hemagglutinins, an elevated peripheral white blood cell count, and the presence of atypical lymphocytes known as *Downey cells*. Patients also may develop antibodies to the capsid antigen of Epstein-Barr (EB) virus as measured by immunofluorescent staining of virus-bearing cells.

71. The answer is A. *(MMWR 37:737–738, 1988.)* HTLV-I infection is highly cell-associated. All the components listed except plasma could carry cells infected with HTLV-I. No transmission by plasma fractions has been documented from units contaminated with HTLV-I.

72. The answer is C. *(Davis, 4/e. pp 900–901.)* Viral cytopathic effects are thought to include a change in the host cell's macromolecular synthesis and the structure of the cell membrane. Viruses may produce cytopathic changes without forming infectious virions and without replicating infectious virus, although the cytopathology is usually fatal to the cell. A particular cytopathic effect is not necessarily associated with a specific virus.

73. The answer is D. *(Balows, Laboratory Diagnosis, pp 235–237.)* With an acute case of primary infection by Epstein-Barr virus (EBV), such as infectious mononucleosis, IgM and IgG antibodies to VCA should be present. Antibodies to EBNA should be absent as they usually appear 2 to 3 months after onset of illness. Culture is not clinically useful because it (1) requires freshly fractionated cord blood lymphocytes, (2) takes 3 to 4 weeks for completion, and (3) is reactive in the majority of seropositive patients.

74. The answer is A. *(Jawetz, 19/e. pp 452–467.)* The *e* antigen seems to be related to the Dane particle, which is presumed to be the intact hepatitis B virus. Possession of the *e* antigen suggests active disease and, thus, an increased risk of transmission of hepatitis to others. HBsAg and *e* antigen are components of hepatitis B and are not shared by other hepatitis viruses.

75. The answer is D. *(Balows, Laboratory Diagnosis, pp 797–799.)* While all the viruses listed will cause hepatitis, the best answer is non-A, non-B hepatitis (hepatitis C). Hepatitis A is transmitted by the fecal-oral route. For an infection of hepatitis delta to be established, an active case of hepatitis B is required. All blood units are specifically screened for hepatitis B. Until 1990 there was no test available for non-A, non-B hepatitis; instead, an indirect test of liver function, serum alanine aminotransferase (ALT; previously designated SGPT), was used to screen for non-A, non-B hepatitis. This test had only a 26 percent sensitivity. Approximately 90 percent of cases of hepatitis associated with transfusion are caused by non-A, non-B hepatitis.

76. The answer is A. *(Balows, Laboratory Diagnosis, pp 665, 679–680.)* Both viruses are retroviruses that use the enzyme reverse transcriptase to make proviral DNA for their RNA genome. Both infect T-cell lymphocytes and possess a glycoprotein envelope. HTLV-I is not closely related to HIV-1. It does not cause immunosuppression and its antigens do not cross-react with HIV-1.

77. The answer is D. *(Jawetz, 19/e. pp 506–516.)* Secondary bacterial pneumonia often produces a syndrome that is clinically distinguishable from that of primary viral pneumonia. The patients most often affected are the elderly, particularly those with chronic pulmonary problems. *Streptococcus pneumoniae, Staphylococcus aureus, Haemophilus influenzae,* and, recently, *Streptococcus pyogenes* are the most common organisms involved in secondary infection.

78. The answer is E. *(Davis, 4/e. pp 925, 935, 937–941.)* While the herpesviruses (HSV, CMV, VZV) are all well known for latency, adenovirus can also form a latent infection in the lymphoid tissue. In 50 to 80 percent of surgically removed tonsils or adenoids, adenovirus can be cultured. The virus has also been cultured from mesenteric lymph nodes, and, in rare cases, viral DNA has been detected in peripheral lymphocytes. Recurrent illness usually does not arise from these latent infections; however, activation can occur in the immunosuppressed.

79. The answer is E. *(Davis, 4/e. pp 972–975, 1043, 1101.)* A therapeutic regimen that includes appropriately administered gamma globulin is effective in the treatment of viral hepatitis A and B. Hyperimmune rabies antiserum prolongs the incubation period of rabies and allows the patient more time to mount an immune response to the vaccine. Although it is not a primary form of treatment for patients with poliomyelitis, passive immunization with pooled gamma globulin can offer adequate protection against the disease.

80. The answer is C. *(Balows, Laboratory Diagnosis, pp 414–415.)* St. Louis encephalitis, yellow fever, and dengue fever are caused by flaviviruses. Western equine encephalitis is caused by an alphavirus. Laboratory diagnosis is usually made by demonstration of a fourfold rise in specific antibody titer in paired sera.

81. The answer is A. *(Jawetz, 19/e. pp 435–436.)* Atypical lymphocytes are the hematologic hallmark of infectious mononucleosis with 90 percent or more of the circulating lymphocytes being atypical in some cases. These abnormal lymphocytes are not pathognomonic for infectious mononucleosis. They are also seen in other diseases, including cytomegalovirus infection, viral hepatitis, toxoplasmosis, rubella, mumps, and roseola.

82. The answer is D. *(MMWR 38:5–7, 1989.)* The Western blotting (WB) assay involves separation of HIV-1 proteins by electrophoresis through a polyacrylamide gel. These separated proteins are then transferred from the gel matrix to nitrocellulose paper. The nitrocellulose strip contains a protein fingerprint of the viral proteins. It is then reacted with the patient's serum and any antigen-antibody complexes are visualized by use of an anti-human IgG conjugated to an enzyme that in the presence of substrate will produce a colored band. Interpretive criteria for the WB test are not presently standardized and various groups recognize different patterns as reactive. Regardless of the interpretive criteria, there remains a "gray zone," which is reported as "indeterminate." This is usually seen with a patient who shows antibody response to only one protein, such as the 24,000-dalton *gag* protein. WB is *not* a good screening test because approximately 15 to 20 percent of people who are not at risk and who have a negative enzyme immunoassay screening test will show an indeterminate WB pattern.

83. The answer is A. *(Jawetz, 19/e. p 401.)* Ribavirin is effective to varying degrees against several RNA- and DNA-containing viruses in vitro. It has been approved for aerosol treatment of respiratory syncytial virus infections in infants. Intravenous administration has proved effective in treating Lassa fever.

84. The answer is D. *(Jawetz, 19/e. pp 384, 385.)* Defective viruses require helper activity from another virus for some step in replication or maturation. Hepatitis D virus is a defective virus. It can replicate only in the presence of hepatitis B virus.

85. The answer is D. *(Davis, 4/e. p 1059.)* Dengue (breakbone fever) is caused by a group B togavirus that is transmitted by mosquitoes. The clinical syndrome usually consists of a mild systemic disease characterized by severe joint and muscle pain, headache, fever, lymphadenopathy, and a maculopapular rash. Hemorrhagic dengue, a more severe syndrome, may be prominent during some epidemics; shock and occasionally death result.

Microbiology

86. The answer is A. *(Jawetz, 19/e. pp 482–486.)* At least five enzymes are located in the core of reoviruses. These include RNA-dependent RNA polymerase, a nucleoside triphosphate phosphohydrolase, guanidyltransferase, and two methylases. All are required for the transcription of the viral RNAs into mRNAs.

87. The answer is D. *(Jawetz, 19/e. pp 454–464.)* HCV is a positive-stranded RNA virus, tentatively classified as a flavivirus. About half of HCV patients develop chronic hepatitis. A large number of infections appear among IV drug abusers. About 90 percent of the cases of transfusion-associated hepatitis are thought to be caused by HCV.

88–91. The answers are: 88-E, 89-A, 90-D, 91-B. *(Jawetz, 19/e. pp 458–462.)* Advances in the serodiagnosis of viral hepatitis have been dramatic, and the findings of specific viral antigens have led to further elucidation of the course of infections. The "Australian antigen," discovered in 1960, was first renamed hepatitis-associated antigen (HAA) and then, finally, hepatitis B surface antigen (HBsAg). It appears in the blood early after infection, before onset of acute illness, and persists through early convalescence. HBsAg usually disappears within 4 to 6 months after the start of clinical illness except in the case of chronic carriers.

Hepatitis B e antigen (HBeAg) appears during the early acute phase and disappears before HBsAg is gone, although it may persist in the chronic carrier. Persons who are HBeAg-positive have higher titers of HBV and therefore are at a higher risk of transmitting the disease. HBeAg has a high correlation with DNA polymerase activity.

The hepatitis B core antigen (HBcAg) is found within the nuclei of infected hepatocytes and not generally in the peripheral circulation except as an integral component of the Dane particle. The antibody to this antigen, anti-HBc, is present at the beginning of clinical illness. As long as there is ongoing HBV replication, there will be high titers of anti-HBc. During the early convalescent phase of an HBV infection, anti-HBc may be the only detectable serologic marker ("window phase") if HBsAg is negative and anti-HBsAg has not appeared.

92–96. The answers are: 92-C, 93-D, 94-D, 95-B, 96-A. *(Jawetz, 19/e. pp 402–407.)* Under the appropriate circumstances, inactivated virus vaccines are given for eastern equine encephalitis and rabies. Vaccine for eastern equine encephalitis is derived from chick embryo cell culture; rabies vaccine is made from duck embryo treated with ultraviolet light or phenol. Vaccines containing inactive virus confer a briefer immunity than those with live virus.

In developed countries, it is recommended that the general population be immunized routinely against measles, rubella, mumps, and poliomyelitis.

Virology *Answers* 37

Live attenuated virus vaccines for these diseases are derived from a variety of tissue cultures; mumps and measles vaccines, for instance, are made from chick embryo tissue culture, and poliomyelitis vaccine from tissue culture of monkey kidney cells. Live attenuated virus vaccines are associated with a longer-lasting antibody response than other types of vaccine; however, there is a risk that the virus can revert to a more virulent form once in the body.

Immunization against smallpox entails the use of a vaccine containing live active virus. The sources of smallpox vaccine are calf or sheep lymph and chorioallantois. Routine immunization of U.S. citizens against smallpox is no longer practiced.

97–101. The answers are: 97-D, 98-A, 99-B, 100-C, 101-E. *(Jawetz, 19/e. pp 127, 426–427, 464–467, 514–516.)* The original vaccine for hepatitis B was prepared by purifying hepatitis B surface antigen (HBsAg) from healthy HBsAg-positive carriers and treating it with viral inactivating agents. The second-generation vaccine for hepatitis B is produced by recombinant DNA in yeast cells containing a plasmid into which the gene for HBsAg has been incorporated.

At the present time, an experimental (attenuated virus) vaccine for hepatitis A is undergoing clinical trials in humans. Until this vaccine becomes available, immune human globulin prepared from large pools of normal adult plasma confers passive protection in about 90 percent of those exposed when given within 1 to 2 weeks after exposure.

Influenza usually occurs in successive waves of infection with peak incidences during the winter months. If only minor antigenic drift is expected for the next influenza season, then the most recent strains of A and B viruses representative of the main antigens are included in the vaccine. Influenza vaccine consists of killed viruses.

Live attenuated measles virus vaccine effectively prevents measles. Protection is provided if given before or within 2 days of exposure. Vaccination confers immunity for at least 15 years.

Acyclovir is an analogue of guanosine or deoxyguanosine that strongly inhibits herpes simplex virus (HSV) but has little effect on other DNA viruses. When employed for the treatment of primary genital infection by HSV, both oral and intravenous formulations have reduced viral shedding and shortened the duration of symptoms.

102–106. The answers are: 102-C, 103-D, 104-E, 105-B, 106-A. *(Jawetz, 19/e. pp 413–416, 427–434, 483–486, 564–566.)* Varicella-zoster is a herpesvirus. Chickenpox is a highly contagious disease of childhood that occurs in the late winter and early spring. It is characterized by a generalized vesicular eruption with relatively insignificant systemic manifestations.

Adenovirus has been associated with ARD among newly enlisted military troops. Crowded conditions and strenuous exercise may account for the severe infections seen in this otherwise healthy group.

Papillomavirus is one of two members of the family Papovaviridae, which includes viruses that produce human warts. These viruses are host-specific and produce benign epithelial tumors that vary in location and clinical appearance. The warts usually occur in children and young adults and are limited to the skin and mucous membranes.

Rotavirus is worldwide in distribution and has been implicated as the major etiologic agent of infantile gastroenteritis. Infection with this virus varies in its clinical presentation from asymptomatic infection to a relatively mild diarrhea to a severe and sometimes fatal dehydration. The exact mode of transmission of this infectious agent is not known.

Infectious mononucleosis caused by cytomegalovirus (CMV) is clinically difficult to distinguish from that caused by Epstein-Barr virus. Lymphocytosis is usually present with an abundance of atypical lymphocytes. CMV-induced mononucleosis should be considered in any case of mononucleosis that is heterophil-negative and in patients with fever of unknown origin.

107–111. The answers are: 107-A, 108-C, 109-B, 110-B, 111-A. *(Jawetz, 19/e. pp 451–467.)* Hepatitis A virus (HAV) possesses a single-stranded linear RNA genome while hepatitis B virus (HBV) contains a double-stranded DNA genome. Detection of anti-HAV IgM in a single serum specimen obtained in the acute or convalescent stages is the quickest and most reliable method to diagnose hepatitis A infection. This antibody is usually present at onset of symptoms and may persist 3 to 6 months. Demonstration of hepatitis B surface antigen (HBsAg) in serum is the most common method of diagnosing HBV infection. Other serologic markers helpful in characterizing infection with HBV include hepatitis B surface antibody (anti-HBs), anti-hepatitis B core (anti-HBc), anti-hepatitis B e antigen (anti-HBe), and hepatitis B e antigen (HBeAg). Several epidemiologic studies have demonstrated that immune serum globulin (ISG) can prevent clinical hepatitis A even when given up to 10 days after exposure. Similar studies have shown that ISG was able to decrease the incidence of hepatitis B infection in exposed persons. Purified, noninfectious HBsAg derived from healthy HBsAg carriers has been used as a vaccine for active immunization for HBV infection.

112–115. The answers are: 112-A, 113-C, 114-B, 115-A. *(Jawetz, 19/e. pp 561–566.)* Human papillomaviruses (HPV) are the causative agents of cutaneous warts as well as proliferative squamous lesions of mucosal surfaces. Although most infections by human papillomavirus are benign, some undergo malignant transformation into in situ and invasive squamous cell carcinoma.

Both HPV and polyomavirus have icosahedral capsids and DNA genomes. JC virus, a polyomavirus, was first isolated from the diseased brain of a patient with Hodgkin's lymphoma who was dying of progressive multifocal leukoencephalopathy (PML). This demyelinating disease occurs usually in immunosuppressed persons and is the result of oligodendrocyte infection by JC virus. JC virus has also been isolated from the urine of patients suffering from demyelinating disease. Cryotherapy and laser treatment are the most popular therapies for warts, although surgery may be indicated in some cases. At the present time, there is no effective antiviral therapy for treatment of infection with polyomavirus or HPV.

116–120. The answers are: 116-C, 117-C, 118-C, 119-B, 120-D. *(Jawetz, 19/e. pp 506–523.)* Orthomyxoviruses and paramyxoviruses are RNA viruses that contain a single-stranded RNA genome. The influenza viruses belong to the orthomyxoviruses. They cause acute respiratory tract infections that usually occur in epidemics. Isolated strains of influenza virus are named after the virus type (influenza A, B, or C) as well as the host and location of initial isolation, the year of isolation, and the antigenic designation of the hemagglutinin and neuraminidase. Both the hemagglutinin and neuraminidase are glycoproteins under separate genetic control, and because of this they can and do vary independently. The changes in these antigens are responsible for the antigenic drift characteristic of these viruses. The paramyxoviruses include several important human pathogens (mumps virus, measles virus, respiratory syncytial virus, and parainfluenza virus). Both paramyxoviruses and orthomyxoviruses possess an RNA-dependent RNA polymerase that is a structural component of the virion and produces the initial RNA.

Bacteriology

DIRECTIONS: Each question below contains four or five suggested responses. Select the **one best** response to each question.

121. A woman appears in the emergency room with symptoms of bacterial vaginosis (nonspecific vaginitis). The vaginal discharge is foul-smelling. Culture of the discharge would be expected to reveal

(A) *Staphylococcus aureus*
(B) *Mycoplasma hominis*
(C) *Mobiluncus*
(D) *Ureaplasma urealyticum*
(E) group B *Streptococcus*

122. A patient reported to his physician that he was bitten by a tiny tick 3 to 4 weeks ago. He noted a confluent red rash around the tick bite followed by a flulike illness. Which of the following laboratory tests would be likely at this time to confirm that the patient has Lyme disease?

(A) *Borrelia burgdorferi* specific IgM antibody (EIA)
(B) *B. burgdorferi* specific IgG antibody (EIA)
(C) Western blot analysis of specific IgG antibody response
(D) Culture of blood for *B. burgdorferi*
(E) Cerebrospinal fluid (CSF) protein analysis

123. The following test results were observed in a woman tested in November who reported being in the woods in Pennsylvania during the past summer: IgG antibody (*Borrelia burgdorferi*) 1:1280; IgM antibody (*B. burgdorferi*) negative. Which one of the following courses of action is the LEAST appropriate?

(A) Order tests for syphilis (VDRL, FTA-ABS) because there are cross-reactions reported with *B. burgdorferi*
(B) Ask the patient if she has aching joints
(C) Consider treatment of the patient with an appropriate antibiotic such as tetracycline
(D) Ask the patient if she has had any neurologic problems, such as Bell's palsy
(E) Ignore the results because there is no Lyme disease in Pennsylvania

Questions 124–125

124. At a church supper in Nova Scotia, the following meal was served: baked beans, ham, coleslaw, eclairs, and coffee. Of the 30 people who attended, 4 senior citizens became ill in 3 days; 1 eventually died. Two weeks after attending the church supper, a 19-year-old girl gave birth to a baby who rapidly became ill with meningitis and died in 5 days. Epidemiologic investigation revealed the following percentages of people who consumed the various food items: baked beans 30 percent, ham 80 percent, coleslaw 60 percent, eclairs 100 percent, and coffee 90 percent. All the following statements are true EXCEPT

(A) this is not a case of food poisoning because only 4 people became ill
(B) the death of the baby may be related to the food consumed at the church supper
(C) based on the epidemiologic investigation, no one food item can be implicated as the cause of the disease
(D) additional data on the microbiologic analysis of the food are required
(E) additional epidemiologic data should include the percentage of those who ate a particular food item who became ill

125. Microbiologic analysis revealed no growth in the baked beans, ham, or coffee; many gram-positive beta-hemolytic, short, rod-shaped bacteria in the coleslaw; and rare gram-positive cocci in the eclairs. The most likely cause of this outbreak is

(A) *Staphylococcus aureus*
(B) *Listeria*
(C) *Clostridium perfringens*
(D) *Clostridium botulinum*
(E) nonmicrobiologic

126. A patient with a peptic ulcer was admitted to the hospital and a gastric biopsy was performed. The tissue was cultured on chocolate agar incubated in a microaerophilic environment at 37°C for 5 to 7 days. At 5 days of incubation, colonies appeared on the plate and were curved, gram-negative rods, oxidase-positive. The most likely identity of this organism is

(A) *Campylobacter jejuni*
(B) *Vibrio parahaemolyticus*
(C) *Haemophilus influenzae*
(D) *Helicobacter* (previously *Campylobacter*) *pylori*
(E) *Campylobacter fetus*

127. *Staphylococcus aureus* was isolated from the draining neck wound of a hospitalized patient. The patient was treated with adequate doses of oxacillin and failed to respond. Subsequently, *S. aureus* was again isolated from the wound site. The most appropriate course of action would be to

(A) continue oxacillin treatment for 2 more weeks
(B) request that the laboratory retest the isolate for resistance to oxacillin
(C) change the antibiotic to vancomycin
(D) change the antibiotic to cephalothin
(E) request that both isolates be phage typed

128. *Mycobacterium avium* is a major opportunistic pathogen in AIDS patients. *M. avium* from AIDS patients can be characterized by all the following statements EXCEPT

(A) the majority of *M. avium* isolates from AIDS patients are pigmented
(B) *M. avium* isolates from AIDS patients are of multiple serovars
(C) all isolates from AIDS patients are acid-fast
(D) most isolates from AIDS patients are resistant to isoniazid and streptomycin
(E) *M. avium* can be isolated from the blood of AIDS patients

129. Organisms resistant to degradative lysosomal enzymes and able to survive lysosomal engulfment include which one of the following?

(A) *Legionella pneumophila*
(B) *Mycobacterium tuberculosis*
(C) *Chlamydia trachomatis*
(D) *Staphylococcus aureus*
(E) *Streptococcus pneumoniae*

130. All the following statements concerning the *Mycobacterium avium/intracellulare* (MAI) complex are true EXCEPT

(A) prior to 1980 there were few cases of MAI disease observed
(B) patients with reduced T-cell–mediated immunity are more susceptible to infection with MAI
(C) MAI infection in AIDS patients is a result of direct contact with another person who is infected with acid-fast bacilli
(D) both the Bactec and Isolator systems are useful for isolation of MAI from blood
(E) acid-fast stain and culture of the stool are important diagnostic aids

131. An inhibitor was designed to block a biologic function in *Haemophilus influenzae*. If the goal of the experiment was to reduce the virulence of *H. influenzae*, the most likely target would be

(A) exotoxin liberator
(B) endotoxin assembly
(C) flagella synthesis
(D) capsule formation
(E) IgA protease synthesis

Bacteriology

132. All the following statements about Kawasaki's syndrome (KS) are true EXCEPT

(A) mites found in house dust may be implicated
(B) rickettsia-like bacteria have been observed in skin biopsies from patients with KS
(C) it is most prevalent in Japan and Hawaii
(D) an atypical strain of *Propionibacterium* may be involved
(E) there is evidence of person-to-person transmission

Questions 133–135

A 21-year-old college student complained of malaise, low-grade fever, and a harsh cough, but not of muscle aches and pains. An x-ray revealed a diffuse interstitial pneumonia in the left lobes of the lung. The WBC count was normal. The student had been ill for a week.

133. Based on the information given, the most likely diagnosis is

(A) *Mycoplasma* pneumonia
(B) pneumococcal pneumonia
(C) staphylococcal pneumonia
(D) influenza
(E) legionellosis

134. Based on the information given, which of the following laboratory tests would most rapidly assist you in making the diagnosis?

(A) Cold agglutinins
(B) Viral culture
(C) Complement fixation test
(D) Gram's stain of sputum
(E) Culture of sputum

135. The following laboratory data were available within 2 days: cold agglutinins—negative; complement fixation (CF) (*M. pneumoniae*)—1:64; viral culture—pending, but negative to date; bacterial culture of sputum on blood agar and MacConkey's agar—normal oral flora. In order to confirm the diagnosis, which of the following procedures could be ordered to achieve a specific and sensitive diagnosis?

(A) Culture of the sputum on charcoal yeast extract
(B) A repeat cold agglutinin test
(C) A DNA probe to the 16S ribosomal RNA of an organism lacking a cell wall
(D) A repeat CF test in 5 days
(E) Another viral culture in one week

136. An antral biopsy was performed on a patient with symptoms of gastric ulcers. A Giemsa stain of the tissue revealed curved, rod-shaped bacteria. Culture on enriched chocolate agar, inactivated at 35°C, revealed a gram-negative, comma-shaped bacillus. The most likely identification is

(A) *Campylobacter jejuni*
(B) *Helicobacter pylori*
(C) *Vibrio cholerae*
(D) *Mycobacterium*
(E) *Escherichia coli*

137. All the following statements concerning rheumatic fever (RF) are true EXCEPT
(A) it is characterized by inflammatory lesions that may involve the heart, joints, subcutaneous tissues, and the central nervous system
(B) the pathogenesis is related to the similarity between a streptococcal antigen and a human cardiac antigen
(C) prophylaxis can be provided by benzathine penicillin
(D) it is a complication of group A streptococcal pharyngitis but usually not of streptococcal skin disease
(E) it is very common in developing countries but extremely rare and decreasing in incidence in the U.S.

138. An experimental compound is discovered that prevents the activation of adenyl cyclase and the resulting increase in cyclic AMP. The toxic effects of which of the following bacteria might be prevented with the use of this experimental compound?
(A) *Vibrio cholerae*
(B) *Corynebacterium diphtheriae*
(C) *Pseudomonas*
(D) *Listeria monocytogenes*
(E) *Brucella*

139. The fermentation patterns for four strains of gram-negative cocci are given below (strains C and D grow on plain nutrient agar). Which of these strains is LEAST likely to be pathogenic for humans?

	Acid produced from		
	Maltose	Dextrose	Sucrose
Strain A	+	+	−
Strain B	−	+	−
Strain C	−	−	−
Strain D	+	+	+

(A) Strain A
(B) Strain B
(C) Strain C
(D) Strain D

140. A 19-year-old male student became ill with diarrhea within 48 h of eating a hamburger at a fast-food shop. Initial culture of his bloody stool revealed no *Salmonella, Shigella, Campylobacter,* or *Yersinia*. A filtrate of the stool was put on Vero cells and cytotoxicity occurred within 24 h. The most likely cause of the illness is
(A) *Escherichia coli* 0157/H7
(B) *E. coli* LT toxin
(C) *E. coli* endotoxin
(D) *Vibrio*
(E) *Clostridium difficile*

141. A patient presents with widely distributed skin nodules. One of the nodules is ulcerated and a scraping reveals acid-fast bacilli. The facial appearance of the patient can be described as "leonine." The most likely diagnosis is

(A) tuberculosis
(B) fungal infection
(C) leprosy
(D) actinomycosis
(E) nocardiosis

142. Infection with *Mycobacterium tuberculosis* is a function of both the virulence of the organism and the resistance of the host. Which one of the following lesions is characteristic of the more advanced disease stages?

(A) Monocytic infiltration
(B) An exanthem on the hands and feet
(C) Eosinophilic granulomas
(D) Granulomatous lesions
(E) PMN-rich abscesses

143. Endotoxin produced by gram-negative bacteria can cause all the following EXCEPT

(A) hemorrhagic tissue necrosis
(B) disseminated intravascular coagulation (DIC)
(C) the Shwartzman phenomenon
(D) fever
(E) hemolytic uremic syndrome

144. The class of antibiotics known as the quinolones are bactericidal. Their mode of action on growing bacteria is thought to be

(A) inhibition of DNA gyrase
(B) inactivation of penicillin-binding protein II
(C) inhibition of β-lactamase
(D) prevention of the cross-linking of glycine
(E) inhibition of reverse transcriptase

145. A patient presented with a facial abscess that had expanded into the contiguous skin surface. Sinus tracts to the skin surface were evident, as was a purulent discharge. The pus was carefully examined and yellow granules were observed. An acid-fast smear revealed no acid-fast bacilli. The patient most likely has

(A) actinomycosis
(B) nocardiosis
(C) staphylococcal infection
(D) streptomycosis
(E) mycetoma

146. Relapsing fever, caused by *Borrelia recurrentis*, is characterized by

(A) an incubation period of 6 weeks
(B) recurring bouts of progressively higher temperature
(C) intense headaches
(D) diarrhea
(E) vomiting

147. If a quellung test was done on the following bacterial isolates, which one would you expect to be negative?

(A) *Streptococcus pneumoniae*
(B) *Klebsiella pneumoniae*
(C) *Haemophilus influenzae*
(D) *Corynebacterium diphtheriae*
(E) *Neisseria meningitidis*

148. A sputum sample was brought to the laboratory for analysis. Gram's stain revealed the following: rare epithelial cells, 8 to 10 polymorphonuclear leukocytes per high-power field, and pleomorphic gram-negative rods. As the laboratory consultant, which of the following interpretations should you make?

(A) The sputum specimen is too contaminated by saliva to be useful
(B) There is no evidence of an inflammatory response
(C) The patient has pneumococcal pneumonia
(D) The patient has Vincent's disease
(E) The appearance of the sputum is suggestive of *Haemophilus* pneumonia

149. Cerebrospinal fluid (CSF) drawn from a newborn baby is noted on Gram's stain to contain gram-positive cocci. The following morning, β-hemolytic colonies resembling streptococci are observed on blood agar plates. Additional tests performed on the CSF to confirm the Gram's stain include

(A) *Limulus* test for endotoxin
(B) bacitracin susceptibility
(C) optochin sensitivity
(D) detection of antigen in the CSF with sensitized latex reagents
(E) detection of toxic granulation in the leukocytes

150. An isolate from a wound culture is a gram-negative rod identified as *Bacteroides fragilis*. Anaerobic infection with *B. fragilis* is characterized by

(A) a foul-smelling discharge
(B) a black exudate in the wound
(C) an exquisite susceptibility to penicillin
(D) a heme-pigmented colony formation
(E) severe neurologic symptoms

151. Vaccines against plague and tularemia can be prepared from all the following EXCEPT

(A) avirulent live bacteria
(B) heat-killed suspensions of virulent bacteria
(C) formalin-inactivated suspensions of virulent bacteria
(D) chemical fractions of the causative bacilli
(E) synthetic capsular polysaccharide material

152. Cell envelopes of both gram-positive and gram-negative bacteria are composed of complex macromolecules. Which of the following statements describes both gram-positive and gram-negative cell envelopes?

(A) They contain a significant amount of teichoic acid
(B) They contain all the common amino acids
(C) Their antigenic specificity is determined by the polysaccharide O antigen
(D) They act as a barrier to the extraction of crystal violet–iodine by alcohol
(E) They are a diffusion barrier to large charged molecules

153. All the following statements concerning chancroid are true EXCEPT that

(A) the disease is caused by *Haemophilus ducreyi*
(B) there is a rapid diagnostic screening test available
(C) chancroid is most common in non-white males
(D) relatively few cases (less than 10 percent) occur in women
(E) there have been sizable outbreaks in the United States

154. A 23-year-old dental student has signs and symptoms of *Listeria* meningitis. The most likely source of infection is

(A) food
(B) an insect bite
(C) contact with an infected patient
(D) an air-conditioning cooling tower
(E) HIV-positive sera

155. An agent is introduced into a growing bacterial colony, and cell multiplication ceases. Removal of the agent, however, allows bacterial cell division to resume. This agent would be described as

(A) a disinfectant
(B) a bactericide
(C) an antiseptic
(D) a bacteriostat
(E) a sterilizer

156. Bacillary dysentery is a disease of young children. In underdeveloped countries, shigellosis is of monumental public health importance. The ability to distinguish between the species of the genus *Shigella* may have important epidemiologic and clinical advantages. Species of *Shigella* primarily responsible for endemic diarrheal disease in developing nations include

(A) *S. dysenteriae*
(B) *S. sonnei*
(C) *S. boydii*
(D) *S. flexneri*
(E) *S. uremicus*

157. Relapsing fever is caused by a strain of *Borrelia*. The reason for relapse is

(A) development of antibiotic resistance
(B) infection with another species of *Borrelia*
(C) sequestration of the organism
(D) development of antigenic mutants
(E) organism-directed shutdown of the immune system

158. A 55-year-old man who is being treated for adenocarcinoma of the lung is admitted to a hospital because of a temperature of 38.9°C (102°F), chest pain, and a dry cough. Sputum is collected. Gram's stain of the sputum is unremarkable and culture reveals many small gram-negative rods able to grow only on a charcoal yeast extract agar. This organism most likely is

(A) *Klebsiella pneumoniae*
(B) *Mycoplasma pneumoniae*
(C) *Legionella pneumophila*
(D) *Chlamydia trachomatis*
(E) *Staphylococcus aureus*

159. A patient was hospitalized after an automobile accident. The wounds became infected and the patient was treated with tobramycin, carbenicillin, and clindamycin. Five days after antibiotic therapy was initiated, the patient developed severe diarrhea and pseudomembranous enterocolitis. Antibiotic-associated diarrhea and the more serious pseudomembranous enterocolitis can be caused by

(A) *Clostridium sordellii*
(B) *Clostridium perfringens*
(C) *Clostridium difficile*
(D) *Staphylococcus aureus*
(E) *Bacteroides fragilis*

160. Pseudomembranous enterocolitis can be diagnosed by all the following EXCEPT

(A) isolation of the causative organism
(B) detection of *C. difficile* antigen by immunologic means
(C) detection of fecal toxin B by tissue culture
(D) detection of fecal toxin A by ELISA
(E) detection of toxins A and B in the blood

161. *Yersinia pestis* is the causative agent of plague. The role that humans play in the plague life cycle is

(A) primary host
(B) primary transmission vector
(C) secondary reservoir
(D) accidental intruder in the rat-flea cycle
(E) none of the above

Bacteriology

162. Which of the following tests is the most sensitive and specific for the diagnosis of primary syphilis?

(A) Frei test
(B) Microhemagglutination *Treponema pallidum* (MHA-TP) test
(C) Venereal Disease Research Laboratories (VDRL) test
(D) Automated reagin test
(E) Rapid plasma reagin (RPR) test

163. Pseudomembranous enterocolitis can be treated most effectively by which one of the following antibiotics?

(A) Cephalothin
(B) Vancomycin
(C) Gentamicin
(D) Penicillin
(E) Chloramphenicol

164. Assuming that the average achievable serum level of gentamicin is 6 to 8 µg/mL, which of the following bacteria is susceptible to gentamicin?

(A) *Escherichia coli* with a minimal inhibitory concentration (MIC) of 10 µg/mL
(B) *E. coli* with an MIC of 12 µg/mL
(C) *Klebsiella* with an MIC of 0.25 µg/mL
(D) *Klebsiella* with an MIC of 6.0 µg/mL
(E) *Klebsiella* with an MIC of 20 µg/mL

165. A patient in the intensive care unit had multiple intravenous lines, including a Hickman catheter. The patient became febrile (40.6°C [105°F]) and multiple blood cultures were drawn. Two days later subculture of three blood-culture bottles revealed "diphtheroids." An antibiotic susceptibility test was performed by mistake on these alleged contaminants and revealed that the isolates were only susceptible to vancomycin. The report from the laboratory should read

(A) "diphtheroids—suspect skin contamination"
(B) "*Corynebacterium diphtheriae* isolated"
(C) "unconfirmed *Corynebacterium JK (C. jaekium)* isolated"
(D) "suggest repeat specimen"
(E) "no significant growth"

166. A 9-year-old child is brought to the emergency room with the chief complaint of enlarged, painful axillary lymph nodes. The resident physician also notes a small, inflamed, dime-shaped lesion surrounding what appears to be a small scratch on the forearm. The lymph node is aspirated and some pus is sent to the laboratory for examination. A Warthin-Starry silver impregnation stain reveals many highly pleomorphic rod-shaped bacteria. The most likely diagnosis is

(A) plague
(B) yersiniosis
(C) tuberculosis
(D) cat-scratch disease
(E) brucellosis

167. A child comes to an emergency room because of an infected dog bite. The wound is found to contain small gram-negative rods. The most likely cause of infection is

(A) *Escherichia coli*
(B) *Haemophilus influenzae*
(C) *Pasteurella multocida*
(D) *Brucella canis*
(E) *Klebsiella rhinoscleromatis*

168. A patient complained to his dentist about a draining lesion in his mouth. A Gram's stain of the pus showed a few gram-positive cocci, leukocytes, and many branched gram-positive rods. The most likely cause of the disease is

(A) *Actinomyces israelii*
(B) *Actinomyces viscosus*
(C) *Corynebacterium diphtheriae*
(D) *Propionibacterium acnes*
(E) *Staphylococcus aureus*

169. The quellung test, used for the direct identification of capsule-containing bacteria, would be most useful for diagnosis of disease in

(A) a child without a spleen
(B) a child with impetigo
(C) a 21-year-old man with gonorrhea
(D) an AIDS patient with pneumocystis pneumonia
(E) a 13-year-old boy with rheumatic fever

170. A 21-year-old female college student went to the college infirmary with the chief complaint of some mild burning during urination and increased frequency of urination. No discharge from the urethra was noted. A "clean-catch" urine specimen was collected and sent to the laboratory for analysis. The urinalysis report returned the next day indicated a "moderate number of leukocytes per high-power field." The bacteriology report read "few *E. coli* (1×10^4 cfu [colony-forming units]/mL)—suspect contamination." This patient most likely has

(A) no disease
(B) sterile pyuria
(C) anterior urethral syndrome
(D) renal disease
(E) gonococcal urethritis

171. A man who has a penile chancre appears in a hospital's emergency room. The VDRL test is negative. The most appropriate course of action for the physician in charge would be to

(A) send the patient home untreated
(B) repeat the VDRL test in 10 h
(C) perform dark-field microscopy for treponemes
(D) swab the chancre and culture on Thayer-Martin agar
(E) perform a Gram stain on the chancre fluid

Bacteriology

172. Granulomatous lesions, which are circumscribed nodular reactions to irritating stimuli, are associated with all the following diseases EXCEPT

(A) cat-scratch disease
(B) coccidioidomycosis
(C) tuberculosis
(D) sarcoidosis
(E) salmonellosis

173. Bacteria cause disease in a number of ways. One mechanism of pathogenesis is the secretion of potent protein toxins. All the following diseases are caused by microbial protein toxins EXCEPT

(A) tetanus
(B) botulism
(C) *Shigella* dysentery
(D) diphtheria
(E) disseminated intravascular coagulation

174. Fever of unknown origin in a farmer who raises goats would most likely be caused by which of the following organisms?

(A) *Brucella melitensis*
(B) *Clostridium novyi*
(C) *Treponema pallidum*
(D) *Histoplasma capsulatum*
(E) *Mycobacterium tuberculosis*

175. Cholera is a toxicogenic dysenteric disease common in many parts of the world. In the treatment of patients who have cholera, the use of a drug that inhibits adenyl cyclase would be expected to

(A) kill the patient immediately
(B) eradicate the organism
(C) increase fluid secretion
(D) reduce intestinal motility
(E) block the action of cholera toxin

176. Most infections from coagulase-negative staphylococci are opportunistic and require some alteration of the normal host defenses. The majority of cases of prosthetic heart valve endocarditis are caused by

(A) *Staphylococcus hemolyticus*
(B) *Staphylococcus aureus*
(C) *Staphylococcus epidermidis*
(D) *Staphylococcus saprophyticus*
(E) *Staphylococcus hyicus*

177. A box of ham sandwiches with mayonnaise prepared by a person with a boil on his neck was left out of the refrigerator for the on-call interns. Three doctors became violently ill approximately 2 h after eating the sandwiches. The most likely cause is

(A) *Staphylococcus aureus* enterotoxin
(B) coagulase from *Staphylococcus aureus* in the ham
(C) *Staphylococcus aureus* leukocidin
(D) *Clostridium perfringens* toxin
(E) penicillinase given to inactivate penicillin in the pork

178. A 19-year-old woman reports to the college infirmary with the chief complaint of frequency of urination and a burning sensation. She has some mild lower back pain. A urine sample is sent to the laboratory, where it is plated on blood agar (BAP) and MacConkey's (MAC) agar. The following day, there are approximately 30,000 cfu/mL on the BAP, nothing on the MAC. Gram's stain of the colonies shows gram-positive cocci in clumps. What is the most likely cause of the patient's urinary tract infection?

(A) *Staphylococcus aureus*
(B) *Staphylococcus epidermidis*
(C) *Staphylococcus saprophyticus*
(D) Group B streptococcus
(E) Pneumococcus

179. *Staphylococcus aureus* causes a wide variety of infections, ranging from wound infection to pneumonia. Treatment of *S. aureus* infection with penicillin is often complicated by the

(A) inability of penicillin to penetrate the membrane of *S. aureus*
(B) production of penicillinase by *S. aureus*
(C) production of penicillin acetylase by *S. aureus*
(D) lack of penicillin binding sites on *S. aureus*
(E) allergic reaction caused by staphylococcal protein

180. Bacterial meningitis in children between the ages of 6 months and 2 years most commonly is caused by

(A) *Neisseria meningitidis*
(B) *Haemophilus influenzae*
(C) *Streptococcus pyogenes*
(D) *Streptococcus pneumoniae*
(E) *Klebsiella pneumoniae*

181. In staphylococci, antibiotic resistance genes can exist either on plasmids or chromosomes. The genes are carried by

(A) prophage
(B) free DNA
(C) transposons
(D) protein A
(E) coagulase

182. Symptoms of *Clostridium botulinum* food poisoning include double vision, inability to speak, and respiratory paralysis. These symptoms are consistent with

(A) invasion of the gut epithelium by *C. botulinum*
(B) secretion of an enterotoxin
(C) endotoxin shock
(D) ingestion of a neurotoxin
(E) activation of cyclic AMP

183. In people who have sickle cell anemia, osteomyelitis usually is associated with which of the following organisms?

(A) *Micrococcus*
(B) *Escherichia*
(C) *Pseudomonas*
(D) *Salmonella*
(E) *Streptococcus*

Bacteriology

184. Group A β-hemolytic streptococci cause both skin infection and pharyngitis. Which of the following statements is true of streptococcal skin infection?

(A) Common sequelae include rheumatic fever
(B) Common sequelae include acute glomerulonephritis
(C) Most streptococcal skin infections cause a rise in antistreptolysin O (ASO) titer
(D) Clinically, infection produces deep, suppurating ulcers
(E) The causative organism cannot be cultured readily

185. In infants, pneumothorax, pneumatocele, and empyema are frequent complications of pneumonia caused by

(A) *Haemophilus*
(B) *Staphylococcus*
(C) *Klebsiella*
(D) *Mycoplasma*
(E) *Streptococcus*

186. A hyperemic edema of the larynx and epiglottis that rapidly leads to respiratory obstruction in young children is most likely to be caused by

(A) *Klebsiella pneumoniae*
(B) *Mycoplasma pneumoniae*
(C) *Neisseria meningitidis*
(D) *Haemophilus influenzae*
(E) *Haemophilus hemolyticus*

187. A survey of 100 seemingly healthy people revealed that 10 harbored *Neisseria meningitidis* in their nasopharynx. The most likely conclusion from the survey is that

(A) many persons have subclinical *Neisseria* pharyngitis
(B) an epidemic of neisserial meningitis is likely
(C) the laboratory was in error and the isolates were probably *N. sicca*
(D) approximately 10 percent of normal, healthy persons have *N. meningitidis*
(E) approximately 90 percent of normal, healthy persons have protective antibodies against *N. meningitidis*

188. Acute hematogenous osteomyelitis is often diagnosed by isolation of the organism from the blood and is caused most often by

(A) *Proteus mirabilis*
(B) *Streptococcus faecalis*
(C) *Staphylococcus epidermidis*
(D) *Staphylococcus aureus*
(E) *Escherichia coli*

189. Diphtheria toxin is produced only by those strains of *Corynebacterium diphtheriae* that are

(A) glucose fermenters
(B) sucrose fermenters
(C) lysogenic for β-prophage
(D) of the mitis strain
(E) encapsulated

190. *Pseudomonas aeruginosa* is a ubiquitous bacterium. The underlying condition of the patient is a major factor in the virulence of *P. aeruginosa*. Which of the following is a major determinant of the pathogenicity of this organism?

(A) Fluorescein
(B) Pyoverdin
(C) Pyocyanin
(D) Phospholipase
(E) Exotoxin A

Questions 191–193

A 28-year-old menstruating woman appeared in the emergency room with the following signs and symptoms: fever, 104°F (40°C); WBC, 16,000/mm^3; blood pressure, 90/65 mmHg; a scarlatiniform rash on her trunk, palms, and soles; extreme fatigue; vomiting; and diarrhea.

191. The patient described in the case above most likely has

(A) scalded skin syndrome
(B) toxic shock syndrome
(C) Guillain-Barré syndrome
(D) chickenpox
(E) staphylococcal food poisoning

192. Culture of the menstrual fluid in the case cited would most likely reveal a predominance of

(A) *Staphylococcus aureus*
(B) *Staphylococcus epidermidis*
(C) *Clostridium perfringens*
(D) *Clostridium difficile*
(E) *Gardnerella vaginalis*

193. The most characteristic finding not yet revealed in the case just presented would be

(A) travel to Vermont
(B) recent exposure to rubella
(C) a retained tampon
(D) heavy menstrual flow
(E) a meal of chicken in a fast-food restaurant

Questions 194–197

A new latex agglutination (LA) reagent for *Haemophilus influenzae* polysaccharide antigen in cerebrospinal fluid was evaluated. Results were compared with the isolation of *H. influenzae* from the CSF. Results were as follows:

LA POS, CULT POS: 25
LA POS, CULT NEG: 5
LA NEG, CULT POS: 5
LA NEG, CULT NEG: 95

194. The sensitivity of LA is

(A) 0 percent
(B) 30 percent
(C) 85 percent
(D) 95 percent
(E) 100 percent

195. The specificity of LA is

(A) 0 percent
(B) 30 percent
(C) 80 percent
(D) 95 percent
(E) 100 percent

196. The negative predictive value of LA is

(A) 10 percent
(B) 80 percent
(C) 95 percent
(D) 110 percent
(E) not calculable

197. The incidence of *H. influenzae* meningitis in the general population is less than 1 percent. If during an epidemic the incidence rose to 3 percent, the negative predictive value of the LA test would

(A) increase
(B) decrease
(C) remain the same
(D) be impossible to calculate
(E) vary as a function of the specificity of the LA test

Microbiology

DIRECTIONS: Each group of questions below consists of lettered headings followed by a set of numbered items. For each numbered item select the **one** lettered heading with which it is **most** closely associated. Each lettered heading may be used **once, more than once, or not at all.**

Questions 198–200

For each situation listed below, choose the most clinically appropriate action.

(A) Do no further clinical workup
(B) Suggest to the laboratory that low colony counts in symptomatic patients may reflect infection
(C) Determine if fluorescent antibody microscopy is available for diagnosing actinomycosis
(D) Consider vancomycin as an alternative antibiotic
(E) Suggest a repeat antibiotic susceptibility test

198. An enterococcus (*Enterococcus faecium*) was isolated from a urine specimen (100,000 cfu/mL). Treatment of the patient with ampicillin and gentamicin failed

199. A patient with symptoms of urinary tract infection had a culture taken, which grew 5×10^3 *E. coli*. The laboratory reported it as "insignificant"

200. A patient appeared in the emergency room with a submandibular mass. A smear was made of the drainage and a bewildering variety of bacteria were seen, including branched, gram-positive rods

Questions 201–205

For each listed disease, choose the antibiotic therapy of choice.

(A) Penicillin
(B) Ampicillin
(C) Erythromycin
(D) Vancomycin
(E) Ceftriaxone

201. Legionellosis

202. Pneumococcal pneumonia

203. Lyme disease

204. Streptococcal pharyngitis

205. Pseudomembranous enterocolitis

Bacteriology

Questions 206–209

Although cholera, a *Vibrio* infection, has been rarely seen in the United States, there have been recent outbreaks of classic cholera associated with shellfish harvested from the Gulf of Mexico. Vibrios are shaped like curved rods, and infections more common than cholera may be caused by a variety of curved-rod bacteria. For each curved-rod bacterium listed below, select the best descriptive phrase.

(A) "String-test"-positive isolate; three serotypes—Ogawa (AB), Inaba (AC), Hikojima (ABC)
(B) Human pathogen, halophilic, lactose-positive; produces heat-labile, extracellular toxin
(C) Human pathogen, halophilic, lactose-negative, sucrose-negative; causes gastrointestinal diseases primarily from ingestion of cooked seafood
(D) Cause of gastroenteritis; reservoir in birds and mammals, optimal growth at 42°C
(E) None of the above

206. *Campylobacter jejuni*

207. *Vibrio cholerae*

208. *Vibrio parahaemolyticus*

209. *Vibrio vulnificus*

Questions 210–213

Each bacterium listed below is a small, gram-negative rod. For each organism, choose the description with which it is most likely to be associated.

(A) Commonly inhabits the canine respiratory tract and is an occasional pathogen for humans; strongly urease-positive
(B) Pits agar, grows both in carbon dioxide and under anaerobic conditions, and is part of the normal oral-cavity flora
(C) Typically infects cattle, requires 5 to 10% carbon dioxide for growth, and is inhibited by the dye thionine
(D) Typically is found in infected animal bites in humans and can cause hemorrhagic septicemia in animals
(E) Manifests different biochemical and physiologic characteristics, depending on growth temperature, and causes a spectrum of human disease, most commonly mesenteric lymphadenitis

210. *Yersinia enterocolitica*

211. *Brucella abortus*

212. *Bordetella bronchiseptica*

213. *Pasteurella multocida*

Microbiology

Questions 214–218

For each bacterium listed below, choose the preferred isolation medium.

(A) Sheep blood agar
(B) Löffler's medium
(C) Thayer-Martin agar
(D) Thiosulfate citrate bile salts sucrose medium
(E) Löwenstein-Jensen agar

214. *Neisseria gonorrhoeae*

215. *Vibrio cholerae*

216. *Mycobacterium tuberculosis*

217. *Corynebacterium diphtheriae*

218. *Staphylococcus aureus*

Questions 219–222

For each of the following body sites, choose that bacterium that is the predominant normal flora (indigenous organism).

(A) α-Hemolytic streptococci
(B) *Lactobacillus*
(C) *Staphylococcus epidermidis*
(D) *Escherichia coli*
(E) *Bacteroides fragilis*

219. Skin

220. Mouth

221. Bowel

222. Vagina

Questions 223–226

For each bacterium below, choose the description with which it is most likely to be associated.

(A) A facultative anaerobe that often inhabits the buccal mucosa early in a neonate's life and can cause bacterial endocarditis
(B) A β-hemolytic organism that causes a diffuse, rapidly spreading cellulitis
(C) An anaerobic, filamentous bacterium that often causes cervicofacial osteomyelitis
(D) A facultative anaerobe that is highly cariogenic and sticks to teeth by synthesis of a dextran
(E) A facultatively anaerobic, rod-shaped bacterium that sticks to teeth and is cariogenic

223. *Streptococcus mutans*

224. *Streptococcus salivarius*

225. *Actinomyces israelii*

226. *Actinomyces viscosus*

Bacteriology

Questions 227–233

Choose the lettered response that best matches each numbered bacterium.

(A) Secretes erythrogenic toxin that causes the characteristic signs of scarlet fever
(B) Produces toxin that blocks protein synthesis in an infected cell and carries a lytic bacteriophage that produces the genetic information for toxin production
(C) Produces at least one protein toxin consisting of two subunits, A and B, that cause severe spasmodic cough usually in children
(D) Requires cysteine for growth
(E) Secretes exotoxin that has been called "verotoxin" and "Shiga-like toxin"; infection is mediated by specific attachment to mucosal membranes
(F) Possesses N-acetylneuraminic acid capsule and adheres to specific tissues by pili found on the bacterial cell surface
(G) Has capsule of polyglutamic acid, which is toxic when injected into rabbits
(H) Synthesizes protein toxin as a result of colonization of vaginal tampons SA
(I) Causes spontaneous abortion and has tropism for placental tissue due to the presence of erythritol in allantoic and amniotic fluid
(J) Secretes two toxins, A and B, in large bowel during antibiotic therapy GDT
(K) Has 82 polysaccharide capsular types; capsule is antiphagocytic; type 3 capsule (β-D-glucuronic acid polymer) most commonly seen in infected adults

227. *Corynebacterium diphtheriae*

228. *Bordetella pertussis*

229. *Francisella tularensis*

230. *Escherichia coli* 0157/H7

231. *Streptococcus pyogenes*

232. *Neisseria meningitidis*

233. *Brucella*

Bacteriology
Answers

121. The answer is C. *(Jawetz, 19/e. p 265.)* Nonspecific vaginoses are rather common. While other organisms such as *Candida albicans* may cause similar symptoms, nonspecific vaginosis is an easily recognizable syndrome. It is characterized by a foul-smelling discharge probably caused by amines produced by anaerobic bacteria such as *Mobiluncus*, a newly recognized genus. *Mobiluncus* and *Gardnerella* may act synergistically. In addition, *Mycoplasma hominis*, *Ureaplasma urealyticum*, *Bacteroides bivius*, *B. capillosis*, *Peptococcus*, and *Eubacterium* are also found occasionally, but their role in vaginosis is unclear.

122. The answer is A. *(Davis, 4/e. p 682.)* If this patient lived in an endemic area for Lyme disease, the clinical presentation would be sufficient for diagnosis. However, laboratory confirmation would be facilitated by detection of specific antibody. Three to four weeks after a tick bite is still early for an increase in IgG antibodies either measured by ELISA or Western blot. IgM response may also be delayed, but appearance of specific IgM antibodies at 4 weeks after a tick bite is not uncommon. A more sensitive technique, Western blotting, would likely be positive with a band at P 41, the flagellar antigen of *B. burgdorferi*, but only if a specific IgM conjugate is used. Culture of the organism is highly unlikely and should not be routinely attempted.

123. The answer is E. *(Jawetz, 19/e. p 285.)* At the present time, Lyme disease may be diagnosed clinically and serologically. Patients who are from endemic areas such as Eastern Pennsylvania and report joint pain and swelling months subsequent to exposure to ticks must be evaluated for Lyme disease and treated if the test is positive. Patients may also report a variety of neurologic problems such as tingling of the extremities, Bell's palsy, and headache. IgM antibody appears soon after the tick bite (10 days to 3 weeks) and persists for 2 months; IgG appears later in the disease but remains elevated for 1 to 2 years, especially in untreated patients. A significant IgG titer is at least 1:320. Most investigators feel that IgM titers of 1:100 are significant; some investigators say that any IgM titer is significant.

124–125. The answers are: 124-A, 125-B. *(Jawetz, 19/e. pp 191–192.)* There have been a number of outbreaks of food poisoning caused by *Listeria mono-*

cytogenes. Listeria is a common inhabitant of farm animals and can be readily isolated from silage, hay, and barnyard soil. Humans at the extremes of age are most susceptible to *Listeria* infection but only recently has food been implicated as a vehicle. In the outbreak in Nova Scotia, it is likely that the cabbage used for the coleslaw was fertilized with animal droppings and not properly washed prior to consumption. Major *Listeria* outbreaks associated with cheese have been seen in the U.S. and most likely have originated from contaminated milk. Epidemiologic investigation often will provide data on attack rates in such outbreaks. The eventual solution of the problem always lies in a combination of epidemiologic, microbiologic, and clinical information. For example, in the Nova Scotia case, it should not be assumed that the eclairs were the culprit based on the fact that everyone ate them.

126. The answer is D. *(Howard, pp 444–446.)* *Helicobacter (Campylobacter) pylori* was first recognized as a possible cause of gastritis and peptic ulcer by Marshall and Warren in 1984. This organism is readily isolated from gastric biopsies but not from stomach contents. It is similar to *Campylobacter* species and grows on chocolate agar at 37°C in the same microaerophilic environment suitable for *C. jejuni* (Campy-Pak or anaerobic jar [Gas Pak] without the catalyst). *H. pylori*, however, grows more slowly than *C. jejuni*, requiring 5 to 7 days incubation. *C. jejuni* grows optimally at 42°C, not 37°C as does *H. pylori*.

127. The answer is B. *(Howard, pp 240–241.)* Oxacillin- and methicillin-resistant *S. aureus* (MRSA) has been rapidly increasing in incidence. MRSA and methicillin-sensitive *S. aureus* (MSSA) coexist in a heterologous population. Treatment of a patient harboring this heterologous population may provide a selective environment for the MRSA. Prior to changing therapy, the susceptibility of the isolate should be determined. Vancomycin has often been used effectively for MRSA. Even though the laboratory results suggest susceptibility, cephalothin should not be used against MRSA as it will likely be ineffective.

128. The answer is B. *(Jawetz, 19/e. p 277.)* There are some interesting characteristics of *M. avium* from AIDS patients. According to data from the National Jewish Hospital and Research Center in Denver and the Centers for Disease Control, 75 percent of the isolates were serovar 4 and 76 percent produced a deep-yellow pigment. Yellow pigment is *not* a characteristic of most isolates of *M. avium*. The significance of these findings is unknown. Most *M. avium* isolates are resistant to isoniazid and streptomycin but susceptible to clofazimine and ansamycin. In vitro susceptibility testing, however, may not be reliable for *M. avium*. A blood culture is often the most reliable way to diagnose the disease.

62   Microbiology

129. The answer is B. *(Davis, 4/e. pp 502–503.)* Organisms such as *M. tuberculosis* (and *M. leprae*) are resistant to the degradative lysosomal enzymes and, hence, able to survive in lysosomes. *Chlamydia, Legionella, Staphylococcus aureus,* and *Streptococcus pneumoniae* do not survive lysosomal engulfment and normally are not found within lysosomes. Chlamydiae prevent lysosomal-phagosomal fusion and legionellae are sequestered in a ribosome-bound organelle that protects them from enzymatic digestion, while staphylococci and pneumococci are extracellular unless engulfed and inactivated by PMNs.

130. The answer is C. *(Howard, pp 196–197, 493–496.)* The MAI complex is now seen as a common cause of disseminated mycobacterial infection in AIDS patients that results from a decreased T-cell response. The disease originates in the gastrointestinal system and spreads to other organs via the blood, where it reproduces slowly but achieves high numbers. Many patients have MAI in the blood with colony counts of 1000 cfu/mL. Examination of feces by smear and culture may provide an early indication of an MAI.

131. The answer is D. *(Davis, 4/e. pp 617–618.)* The major determinant of virulence in *H. influenzae* is the presence of a capsule. There is no demonstrable exotoxin and the role of endotoxin is unclear. While one would expect that IgA protease would inhibit local immunity, the role of this enzyme in pathogenesis is as yet unclear.

132. The answer is E. *(Howard, p 192.)* Kawasaki's syndrome is a mucocutaneous lymph node disease. It occurs predominantly in children. Clinical features include prolonged fever, inflammation of the eyes and pharynx, and lymphadenopathy. The relationship between house mites and KS is strong enough that young children should not be in the vicinity of a recently shampooed rug.

133–135. The answers are: 133-A, 134-A, 135-C. *(Jawetz, 19/e. pp 268–269.)* *Mycoplasma pneumoniae* causes a respiratory infection known as "primary atypical pneumonia" or "walking pneumonia." Although disease caused by *M. pneumoniae* can be contracted year round, thousands of cases occur during the winter months in all age groups. The disease, if untreated, will persist for 2 weeks or longer. Rare but serious side effects include cardiomyopathies and central nervous system complications. Infection with *M. pneumoniae* may be treated with either erythromycin or tetracycline. The organism lacks a cell wall and so is resistant to the penicillin and the cephalosporin groups of antibiotics.

Until recently, diagnostic tests have been of limited value. Up to 50 percent of cases may *not* show cold agglutinins, an insensitive and nonspecific acute-phase reactant. However, if cold agglutinins are present, a quick diagnosis can be made if signs and symptoms are characteristic. Complement fixation tests that measure an antibody to a glycolipid antigen of *M. pneumoniae* are useful but not routinely performed in most laboratories. Also, cross-reactions may occur. Culture of *M. pneumoniae*, while not technically difficult, may take up to 2 weeks before visible growth is observed. A DNA probe is available. It is an ^{125}I probe for the 16S ribosomal RNA of *M. pneumoniae*. Evaluations in a number of laboratories indicate that compared with culture it is highly sensitive and specific.

136. The answer is B. *(Howard, p 446.)* Helicobacter pylori has been implicated in cases of gastritis and gastric ulcers. While there is not a direct cause-and-effect relationship, it has been observed that most patients with symptoms harbor the organism. Histologic examination of tissue for characteristic bacilli is at least as sensitive as culture for detection of *H. pylori*. Also, patients with gastritis have antibody titers to *H. pylori*.

137. The answer is E. *(Howard, p 256.)* Rheumatic fever (RF) is a disease that causes polyarthritis, carditis, chorea, and erythema marginatum. The mechanism of damage appears to be autoimmune; that is, antibodies are synthesized to a closely related streptococcal antigen such as M-protein, but these same antibodies cross-react with certain cardiac antigens such as myosin. Until recently, RF was very rare in the U.S. In 1986, there were at least 135 cases of RF in Utah. Subsequently, scattered cases of RF have occurred in other states. Epidemiologists do not have a reason for this increase in RF. Some evidence suggests that there may be a genetic predisposition to the disease. Intramuscular injection of benzathine penicillin is effective treatment for and prophylaxis against group A streptococcal infection.

138. The answer is A. *(Davis, 4/e. p 584.)* The toxin of *Vibrio cholerae* and LT enterotoxin from *Escherichia coli* are similar. The B subunits of the toxins bind to ganglioside G_{M1} receptors on the host cell. The A subunits catalyze transfer of the ADP-ribose moiety of ADP to a regulatory protein known as G_s. This activated G_s stimulates adenyl cyclase. Cyclic AMP is increased as is fluid and electrolyte release from the crypt cells into the lumen of the bowel. Watery, profuse diarrhea ensues.

139. The answer is D. *(Jawetz, 19/e. pp 250–255.)* Pathogenic neisseriae (*Neisseria meningitidis, N. gonorrhoeae*) will not grow on plain agar; they grow best on blood-enriched plates in the presence of 10% carbon dioxide.

Branhamella catarrhalis and *N. sicca* grow on plain nutrient agar. *N. meningitidis* (strain A in the question) produces acid from maltose and dextrose, whereas *N. gonorrhoeae* (strain B) ferments only dextrose. Strain C could be either *B. catarrhalis* or *N. flavescens*. *N. sicca* (strain D) produces acid from sucrose, maltose, and dextrose. *B. catarrhalis* is known to be an etiologic agent of pneumonia, while *N. sicca* is normal flora.

140. The answer is A. *(Balows, Clinical Microbiology, 5/e. p 369.)* Food poisoning with *E. coli* 0157/H7 causes hemorrhagic colitis, often seen after eating beef hamburgers. The same organism also causes a hemorrhagic cystitis. The toxin, called *Shiga-like toxin,* can be demonstrated in Vero cells, but the cytotoxicity must be neutralized with specific antiserum. With the exception of sorbitol fermentation, there is nothing biochemically distinctive about these organisms.

141. The answer is C. *(Davis, 4/e. pp 661–663.)* Leprosy is caused by *Mycobacterium leprae,* an acid-fast bacillus. The organism cannot be cultured on artificial media but grows in the foot pads of mice and in armadillos. The lesions are similar to those of tuberculosis. The organism has a predilection for skin and nerves so that nodules may be distributed widely. The "leonine" appearance of the patient is difficult to describe but characteristic of leprosy.

142. The answer is D. *(Davis, 4/e. pp 652–653.)* In the initial stages of infection, an acute inflammatory (exudative) lesion is seen in the lung and consists of polymorphonuclear leukocytes around the bacteria. When the patient becomes hypersensitive to tuberculoprotein, granulomas are formed. The macrophages become arranged in a concentric circle to form the typical "tubercle."

143. The answer is E. *(Jawetz, 19/e. pp 136–138.)* Endotoxins of gram-negative bacteria are heat-stable lipopolysaccharides derived from the cell wall. They are responsible for many of the symptoms and complications of gram-negative infections, including hemorrhagic tissue necrosis, disseminated intravascular coagulation (DIC), leukopenia, and fever. The Shwartzman phenomenon, a complex reaction to experimentally injected endotoxin, resembles DIC. While hemolytic uremic syndrome is mediated by *Escherichia coli,* a gram-negative bacterium, it is likely that the primary lesion is a function of a protein toxin resembling the toxin of *Shigella dysenteriae* (Shiga-like toxin or verotoxin).

144. The answer is A. *(Davis, 4/e. pp 206–207.)* A new class of antibiotics, the quinolones, has one member, nalidixic acid, that has been available for years. The new representatives are much more active biologically and are effective against virtually all gram-negative bacteria and most gram-positive bacteria. They include norfloxacin, ofloxacin, ciprofloxacin, enoxacin, and the fluorinated quinolones such as lomefloxacin. These antibiotics kill bacteria by inhibition of synthesis of nucleic acid, more specifically, DNA gyrase. Resistance to quinolones has been observed and appears to be a class-specific phenomenon. An exception is that when an organism is resistant to nalidixic acid, elevated minimal inhibitory concentrations (MICs) will generally apply to other quinolones, although these MICs will still be within the range of susceptibility.

145. The answer is A. *(Davis, 4/e. pp 666–671.)* *Nocardia*, *Actinomyces*, and *Streptomyces* are actinomycetes, a group of filamentous bacteria that resemble fungi. *Nocardia* is an aerobic, branched, gram-positive, rod-shaped bacterium that is only sometimes capnophilic; *Actinomyces* is an anaerobic, branched, gram-positive, rod-shaped bacterium. *Streptomyces* is filamentous and has spores that may be acid-fast. All three genera are pathogenic. *Nocardia* causes a pulmonary infection resembling tuberculosis, and *Actinomyces* causes cervicofacial osteomyelitic disease. Both *Nocardia* and *Actinomyces* may cause abscesses with either burrowing or communicating sinus tracts. Diseases from *Streptomyces* and *Nocardia* may be indistinguishable. *Actinomyces* is not acid-fast, but *Nocardia* is—i.e., carbol-fuchsin is retained in the cell envelope following acid treatment.

146. The answer is C. *(Jawetz, 19/e. pp 283–285.)* Relapsing fever is characterized by the sudden onset of chills, fever, and severe headache after an incubation period of 3 to 10 days. Splenomegaly and jaundice often occur. The fever ends abruptly in 3 to 4 days but usually recurs 2 to 14 days later. As many as 10 febrile episodes, which become progressively less severe, can occur. Relapses are thought to be due to alterations in the antigenic structure of *Borrelia recurrentis*, the causative organism. Neither diarrhea nor vomiting is a common symptom of relapsing fever.

147. The answer is D. *(Howard, pp 220–221.)* The quellung test determines the presence of bacterial capsules. Specific antibody is mixed with the bacterial suspension or with clinical material. The polysaccharide capsule–antibody complex is visible microscopically. The test is also termed *capsular swelling*. The capsules of *Streptococcus pneumoniae*, *Neisseria meningitidis*, *Haemophilus influenzae*, and *Klebsiella pneumoniae* play a role in the pathogenicity of the organisms. These surface structures inhibit phagocytosis, perhaps by preventing attachment of the leukocyte pseudopod. *Corynebacterium diphtheriae* is nonencapsulated; its pathogenicity is dependent on synthesis and excretion of a protein toxin.

148. The answer is E. *(Howard, pp 210–211.)* Many sputum specimens are cultured unnecessarily. Sputum is often contaminated with saliva or is almost totally made up of saliva. These specimens rarely reveal the cause of the patient's respiratory problem and may provide laboratory information that is harmful. The sputum in question appears to be a good specimen. The pleomorphic gram-negative rods are suggestive of *Haemophilus* but culture of the secretions is necessary.

149. The answer is D. *(Balows, Clinical Microbiology, 5/e. pp 244–248.)* Clinically, the infant described in the question has group B streptococcal meningitis. Although most streptococcal disease in humans is caused by group A streptococci, newborn infants can be infected with group B streptococci, which normally reside in the vagina. After Gram's stain and culture have been performed, laboratory identification should include either a test for hippurate hydrolysis or the CAMP test and confirmation of the serogroup of the organism. The CAMP test determines the presence of a substance that is secreted by group B streptococci and that enhances hemolysis production by *Staphylococcus aureus*; other streptococci rarely can elicit a positive CAMP test. However, a more rapid diagnosis may be made by detecting specific group B streptococcal carbohydrate antigen in the CSF. This latex agglutination procedure is approximately equivalent to the Gram's stain in sensitivity.

150. The answer is A. *(Balows, Clinical Microbiology, 5/e. pp 488–494.)* *Bacteroides fragilis* is a constituent of normal intestinal flora and readily causes wound infections often mixed with aerobic isolates. These anaerobic, gram-negative rods are uniformly resistant to aminoglycosides and usually to penicillin as well. Reliable laboratory identification may require multiple analytical techniques. Generally, wound exudates smell bad owing to production of organic acids by such anaerobes as *B. fragilis*. Black exudates or a black pigment (heme) in the isolated colony is usually a characteristic of *Bacteroides* (*Porphyromonas*) *melaninogenicus*, not *B. fragilis*. Potent neurotoxins are synthesized by the gram-positive anaerobes such as *Clostridium tetani* and *C. botulinum*.

151. The answer is E. *(Jawetz, 19/e. pp 245–249.)* Vaccines against *Yersinia pestis* and *Francisella tularensis* may be commercially prepared from avirulent live bacteria, heat-killed or formalin-inactivated virulent bacteria, and chemical fractions of the bacilli. These vaccines are not specific subunit vaccines to capsular polysaccharide such as pneumococcal and *Haemophilus influenzae* vaccines. They provide some immunity to tularemia and bubonic plague, but not to pneumonic plague. Prophylactic tetracycline therapy provides adequate protection to persons living in areas endemic for plague.

Bacteriology Answers 67

152. The answer is E. *(Jawetz, 19/e. pp 7–31.)* Bacterial cell envelopes consist of both the cell wall and cell membrane; in gram-positive and gram-negative bacteria, this cytoplasmic membrane acts as a diffusion barrier to large charged molecules. The cell envelope of gram-negative bacteria contains lipoprotein, lipopolysaccharide, and peptidoglycan molecules; the polysaccharide component of the lipopolysaccharide is the O antigen, the chief surface antigen of these bacteria. On the other hand, gram-positive organisms contain large amounts of teichoic acids, which are important surface antigens for these bacteria. The cell wall of gram-positive bacteria acts as a barrier to the extraction of crystal violet–iodine complex by alcohol; this property is the basis of the Gram's stain.

153. The answer is B. *(Howard, pp 284–285.)* Chancroid is a sexually transmitted disease caused by *Haemophilus ducreyi*. It is usually associated with poor socioeconomic conditions. While the culture of the organism has improved recently, there still is no reliable diagnostic screening test. The diagnosis is commonly made on clinical grounds.

154. The answer is A. *(Davis, 4/e. pp 720–721.)* *Listeria monocytogenes* is a ubiquitous organism that can be isolated from a wide variety of animals and birds. Infections with *Listeria* are observed most commonly in immunologically compromised hosts and in patients at the extremes of age. However, most recent *Listeria* infections have been due to contaminated milk and cheese and occasionally vegetable products such as coleslaw. Veterinarians who come into contact with infected animals are also at risk. Dentists are at no increased risk.

155. The answer is D. *(Jawetz, 19/e. p 45.)* Antimicrobial agents that inhibit bacterial multiplication but do not kill the cells themselves are called bacteriostats. Cell multiplication resumes once the bacteriostatic agent is removed from the environment. Bactericidal agents, however, do kill bacteria.

156. The answer is D. *(Davis, 4/e. pp 574–576.)* The species of *Shigella* can be distinguished as follows:

Species	Serotype	Mannitol fermentation	Decarboxylation of ornithine
S. dysenteriae	A	–	–
S. flexneri	B	+	–
S. boydii	C	+	–
S. sonnei	D	+	+

S. flexneri is the most common species in underdeveloped nations. *S. sonnei* is responsible for most of the shigellosis in the U.S. *S. dysenteriae* causes a far more serious form of dysentery than do the other three species.

157. The answer is D. *(Davis, 4/e. pp 681–682.)* Relapsing fever can be caused by either *B. recurrentis* in an epidemic form or by other *Borrelia* species, such as *B. hermsii*, as an endemic disease. The term *relapsing fever* aptly describes the disease. At each relapse, antigenically distinct organisms appear, specific antibodies are formed, and the patient feels better—until the next time. Apparently, DNA rearrangements affecting a major surface protein are responsible for the antigen shifts. There are antigenic similarities between *B. hermsii* and *B. burgdorferi*, the causative agent of Lyme disease; however, the clinical characteristics of infection are very different.

158. The answer is C. *(Balows, Clinical Microbiology, 5/e. pp 442–444.)* The symptoms of Legionnaires' disease are similar to those of mycoplasmal pneumonia and influenza. Affected persons are moderately febrile, complain of pleuritic chest pain, and have a dry cough. Unlike *Klebsiella* and *Staphylococcus*, *Legionella pneumophila* exhibits fastidious growth requirements. Charcoal yeast extract agar either with or without antibiotics is the preferred isolation medium. While sputum may not be the specimen of choice for *Legionella*, the discovery of small gram-negative rods by direct fluorescent antibody (FA) technique should certainly heighten suspicion of the disease. *L. pneumophila* is a facultative intracellular pathogen and enters macrophages without activating their oxidizing capabilities. The organisms bind to macrophage C receptors, which promote engulfment.

159. The answer is C. *(Howard, pp 408–409.)* Patients treated with antibiotics develop diarrhea that, in most cases, is self-limiting. However, in some instances, particularly in those patients treated with ampicillin or clindamycin, a severe, life-threatening pseudomembranous enterocolitis develops. This disease has characteristic histopathology, and membranous plaques can be seen in the colon by endoscopy. Pseudomembranous enterocolitis and antibiotic-associated diarrhea are caused by an anaerobic gram-positive rod, *Clostridium difficile*. It has been recently shown that *C. difficile* produces a protein toxin with a molecular weight of about 250,000. The "toxin" is, in fact, two toxins, toxin A and toxin B. Both toxins are always present in fecal samples, but there is approximately one thousand times more toxin B than toxin A. Toxin A has enterotoxic activity—that is, it elicits a positive fluid response in ligated rabbit ileal loops—whereas toxin B appears to be primarily a cytotoxin.

Bacteriology Answers 69

160. The answer is E. *(Howard, pp 408–409.)* *C. difficile* toxin B is detected by adding fecal filtrates to cell cultures. Many cell cultures respond to this toxin with characteristic cytopathology. Alternatively, cell-associated antigen can be detected immunologically by latex agglutination. An ELISA test is now available for toxin A and also one for both toxins A and B in the feces. In yet another diagnostic method, feces can be cultured and *C. difficile* isolated. Toxin assays should be done on the isolates because not all *C. difficile* bacteria produce toxin. *C. difficile* toxin is not detected in the blood.

161. The answer is D. *(Davis, 4/e. pp 602–603.)* Rats are the primary reservoir of plague. They usually die of the disease but occasionally develop a chronic form. Fleas infesting the rat then transmit *Y. pestis* to another rat. If no rat is available, the flea may transmit the disease to humans. Humans, then, are accidental hosts and do not serve as a reservoir for the disease.

162. The answer is B. *(Howard, pp 506–509.)* Serologic tests for syphilis, which is caused by the spirochete *Treponema pallidum*, fall into two groups. Tests that are sensitive but not specific for syphilis include the Venereal Disease Research Laboratories (VDRL), rapid plasma reagin (RPR), and automated reagin tests, which detect the presence of reagin, an antibody-like compound produced during spirochetal infection. Tests both sensitive and specific detect the presence of specific antitreponemal antibodies and include the fluorescent treponemal antibody-absorption (FTA-ABS) test, the treponemal immobilization (TPI) test, and the microhemagglutination *Treponema pallidum* (MHA-TP) test. The Frei skin test, used for the diagnosis of lymphogranuloma venereum, a chlamydial disease, is no longer available. Because the MHA-TP, FTA-ABS, and TPI tests are labor-sensitive, the nontreponemal tests are much more suitable for screening, even though they may be less specific (i.e., false positives are common). The use of PCR amplified DNA probe for *T. pallidum* may prove to be the most definitive test for syphilis, particularly CNS syphilis.

163. The answer is B. *(Howard, pp 408–409.)* It appears that *C. difficile* colonization occurs when the normal gastrointestinal flora has been altered. Relatively few adults (less than 4 percent) carry *C. difficile* as normal flora. However, up to 60 percent of infants may harbor *C. difficile* with no apparent disease. Vancomycin and metronidazole are antibiotics especially effective in the treatment of pseudomembranous enterocolitis.

164. The answer is C. *(Balows, Clinical Microbiology, 5/e. pp 1105–1107.)* The interpretation of quantitative antimicrobial susceptibility tests is based on both the minimal inhibitory concentration (MIC) and the achievable blood level of a given antibiotic. An MIC greater than the achievable blood concentration of an antibiotic suggests resistance. An MIC at or near the achievable level is equivocal, and an MIC significantly lower than the achievable level—say, by 75 percent—suggests susceptibility of the isolate to the antibiotic being tested. Thus, *Klebsiella* listed in the question as having an MIC of 0.25 μg/mL is susceptible to gentamicin with an average serum level of 6 to 8 μg/mL.

165. The answer is C. *(Howard, p 424.)* Reports have indicated that a hitherto unrecognized corynebacterium, *Corynebacterium JK (C. jaekium)*, causes significant morbidity in hospitalized patients, especially those with intravenous catheters. *Corynebacterium JK* can be distinguished from other members of the genus by biochemical and cultural tests. Another common characteristic of *JK* is its antimicrobial resistance pattern. *JK* is commonly susceptible only to vancomycin.

166. The answer is D. *(Mandell, 3/e. pp 1381–1382.)* While the essential information (i.e., the evidence that the child in question was scratched by her pet cat) is missing, the clinical presentation points to a number of diseases, including cat-scratch disease (CSD). Until recently the etiologic agent of CSD was unknown. Recent evidence indicates that it is a pleomorphic rod-shaped bacterium that has been named *Afipia*. It is best demonstrated in the affected lymph node by a silver impregnation stain. One reason for the repeated failure to isolate this organism is that it apparently loses its cell wall and is very sensitive to the artificial conditions of culture. New information indicates that *Afipia* will grow on artificial media designed for the isolation of *Legionella*.

167. The answer is C. *(Jawetz, 19/e. p 248.)* *Pasteurella multocida*, a coccobacillary gram-negative rod, is part of the normal mouth flora of dogs and cats. Consequently, many animal bites become infected with this microorganism. It is susceptible to penicillin, although multiresistant strains have been recovered from pigs and sheep. *P. multocida* has four different capsular types–designated A, B, D, and E—that correlate with disease production and host predilection; however, serotyping of these isolates is beyond the resources of most laboratories.

168. The answer is A. *(Balows, Clinical Microbiology, 5/e. pp 530–531.)* The patient presented with typical symptoms of actinomycosis. *Actinomyces israelii* is normal flora in the mouth. However, it causes a chronic draining infection, often around the maxilla or the mandible, with osteomyelitic changes. Treatment is high-dose penicillin for 4 to 6 weeks. The diagnosis of actinomycosis is often complicated by the failure of *A. israelii* to grow from

Bacteriology Answers 71

the clinical specimen. It is an obligate anaerobe. Fluorescent antibody (FA) reagents are available for direct staining of *A. israelii*. A rapid diagnosis can be made from the pus. FA conjugates are also available for *A. viscosus* and *A. odontolyticus*, anaerobic actinomycetes that are rarely involved in actinomycotic abscesses.

169. The answer is A. *(Howard, pp 253–259.)* When antiserum to capsular polysaccharide is added to a slide preparation of organisms, microscopically the capsule appears swollen. This phenomenon, the quellung reaction, can be used for rapid identification of pneumococci (*S. pneumoniae*). Whether the capsule actually "swells" is questionable; however, visualization of the capsule is improved noticeably following the specific antigen-antibody reaction. Such a test would be most helpful in asplenic patients who are prone to pneumococcal disease. The test would not be useful for patients with other streptococcal infections (impetigo, rheumatic fever), infection with *Neisseria gonorrhoeae*, or opportunistic pulmonary infections such as pneumocystis pneumonia (PCP).

170. The answer is C. *(Howard, pp 191–192.)* It has recently been reported that many women may suffer from anterior urethral syndrome (AUS), which is a relatively mild form of urinary tract infection. Unfortunately, the methods used for processing of urine for culture may miss the causative agents of AUS. It has been shown that the most sensitive colony-count breakpoint for these symptomatic patients is not 1×10^5 cfu/mL but 1×10^2 cfu/mL. The causative agent is usually *E. coli*, but *Chlamydia* has been isolated also. Without prior arrangements the laboratory may discard the specimens as contaminated.

171. The answer is C. *(Jawetz, 19/e. pp 281–282.)* In men, the appearance of a hard chancre on the penis characteristically indicates syphilis. Even though the chancre does not appear until the infection is 2 or more weeks old, the VDRL test for syphilis still can be negative despite the presence of a chancre (the VDRL test may not become positive for 2 or 3 weeks after initial infection). However, a lesion suspected of being a primary syphilitic ulcer should be examined by dark-field microscopy, which can reveal motile treponemes.

172. The answer is E. *(Howard, pp 31–32.)* Granulomatous lesions are circumscribed nodular lesions characterized by the presence of macrophages. They may persist for a long time as sites of smoldering inflammation. Granulomas develop in a variety of diseases, including tuberculosis, sarcoidosis, coccidioidomycosis, and cat-scratch disease. Chemical and mineral irritation also can produce granulomatous reactions.

Microbiology

173. The answer is E. *(Howard, pp 28–29.)* Protein exotoxins are diffusible substances elaborated chiefly by gram-positive organisms, whereas lipopolysaccharide endotoxins are cell-wall components of certain gram-negative bacteria. The exotoxins of *Clostridium tetani* and *C. botulinum* act directly on the nervous system. The "Shiga toxin" of *Shigella* dysentery acts on the smaller cerebral blood vessels. Diphtheria exotoxin affects body cells in general. Disseminated intravascular coagulation (DIC) results from many conditions, including the action of gram-negative bacterial endotoxin on the intrinsic clotting system.

174. The answer is A. *(Jawetz, 19/e. pp 241–243.)* Brucellae are small, aerobic, gram-negative coccobacilli. Of the four well-characterized species of *Brucella*, only one—*B. melitensis*—characteristically infects both goats and humans. Brucellosis may be associated with gastrointestinal and neurologic symptoms, lymphadenopathy, splenomegaly, hepatitis, and osteomyelitis.

175. The answer is E. *(Jawetz, 19/e. pp 230–232.)* Cholera is a toxicosis. The mode of action of cholera toxin is to stimulate the activity of adenyl cyclase, an enzyme that converts ATP to cyclic AMP. Cyclic AMP stimulates the secretion of chloride ion, and affected patients lose copious amounts of fluid. A drug that inhibits adenyl cyclase thus might block the effect of cholera toxin.

176. The answer is C. *(Howard, pp 238–239.)* The widespread use of vascular access routes and prosthetic devices has led to a dramatic increase in the clinical importance of coagulase-negative staphylococci, notably *S. epidermidis*. This organism is responsible for 1 to 10 percent of cases of native valve endocarditis and 40 to 80 percent of cases of prosthetic valve endocarditis. Mortality in infected patients ranges from 63 to 74 percent.

177. The answer is A. *(Howard, pp 237–240.)* Certain strains of staphylococci elaborate an enterotoxin that is frequently responsible for food poisoning. Typically, the toxin is produced when staphylococci grow on foods rich in carbohydrates and is present in the food when it is consumed. The resulting gastroenteritis is dependent only on the ingestion of toxin and not on bacterial multiplication in the gastrointestinal tract. Characteristic symptoms are nausea, vomiting, abdominal cramps, and explosive diarrhea. The illness rarely lasts more than 24 h.

178. The answer is C. *(Jawetz, 19/e. pp 194–199.)* *Staphylococcus saprophyticus* is a common cause of urinary tract infections in young women. Staphylococci, even though they are present in numbers less than 1×10^5 cfu/mL, should be identified. *S. epidermidis* is usually found as a skin contaminant in urine but is a common cause of infection from intravascular catheters and prostheses. The following table differentiates the three most common species of *Staphylococcus*:

	Coagulase	Mannitol	DNase	Novobiocin*
S. aureus	+	+	+	Susceptible
S. epidermidis	−	−	−	Susceptible
S. saprophyticus	−	−	−	Resistant

*Novobiocin is for diagnostic purposes only.

179. The answer is B. *(Jawetz, 19/e. pp 194–199.)* Staphylococci are gram-positive, non-spore-forming cocci. Clinically, their antibiotic resistance poses major problems. Many strains produce β-lactamase (penicillinase), an enzyme that destroys penicillin by opening the lactam ring. Drug resistance, mediated by plasmids, may be transferred by transduction.

180. The answer is B. *(Howard, pp 273, 279–286.)* Except during a meningococcal epidemic, *Haemophilus influenzae* is the most common cause of bacterial meningitis in children. The organism is occasionally found to be associated with respiratory tract infections or otitis media. *H. influenzae*, *Neisseria meningitidis*, and *Streptococcus pneumoniae* account for 80 to 90 percent of all cases of bacterial meningitis. A purified polysaccharide vaccine for *H. influenzae* type B has been available for a number of years; however, it is relatively ineffective, particularly in young children. A recently tested conjugate vaccine (protein and polysaccharide) shows more promise.

181. The answer is C. *(Davis, 4/e. pp 543–544.)* Particularly in *Staphylococcus aureus*, resistance genes are carried on transposons. Determinants of resistance to erythromycin, spectinomycin, penicillin, gentamicin, and methicillin have been noted. Toxic shock syndrome toxin (TSST-1) is also carried on a transposon.

182. The answer is D. *(Jawetz, 19/e. pp 183–184.)* Clostridium botulinum growing in food produces a potent neurotoxin that causes diplopia, dysphagia, respiratory paralysis, and speech difficulties when ingested by humans. The toxin is thought to act by blocking the action of acetylcholine at neuromuscular junctions. Botulism is associated with high mortality; fortunately, *C. botulinum* infection in humans is rare.

183. The answer is D. *(Howard, p 323.)* Many types of infection, notably respiratory tract infections and osteomyelitis, are common in people who have sickle cell anemia. For unknown reasons, *Salmonella* is implicated frequently in these infections. Osteomyelitis in other persons is caused most often by *Staphylococcus*.

184. The answer is B. *(Balows, Clinical Microbiology, 5/e. pp 3–10.)* Skin streptococci are usually nephritogenic and not rheumatogenic. Not uncommonly, patients who have acute glomerulonephritis do not show an elevated titer of antistreptolysin O (ASO) but do have antibody titers to other streptococcal antigens, such as diphosphopyridine nucleotidase and deoxyribonuclease (DNase). Streptococcal skin infection (impetigo) most often affects young children, is highly contagious, and produces superficial blisters.

185. The answer is B. *(Jawetz, 19/e. pp 196–197.)* During the clinical course of primary staphylococcal pneumonia in infants, abscess formation and necrosis can occur throughout the lung parenchyma. These abscesses can rupture into bronchial walls or the pleural cavity, producing pyopneumothorax or pneumatoceles. Surgical intervention often is required.

186. The answer is D. *(Jawetz, 19/e. pp 237–239.)* *Haemophilus influenzae* is a gram-negative bacillus. In young children it can cause pneumonitis, sinusitis, otitis, and meningitis. Occasionally, it produces a fulminative laryngotracheitis with such severe swelling of the epiglottis that tracheostomy becomes necessary. Clinical infections with this organism after the age of 3 years are less frequent.

187. The answer is D. *(Jawetz, 19/e. pp 254–255.)* Approximately 10 percent of healthy adults are carriers of *Neisseria meningitidis*. The percentage increases when people are housed in close quarters, such as military barracks. Carriage of the organism usually predisposes to the formation of protective antibody. However, the presence of a nasopharyngeal source of *N. meningitidis* in closely associating people who do not have protective antibodies may often lead to major outbreaks of meningococcal disease. There is a commercially available vaccine for *N. meningitidis* groups A and C.

188. The answer is D. *(Jawetz, 19/e. pp 196–197.)* *Staphylococcus aureus* is implicated in the majority of cases of acute osteomyelitis, which affects children most often. A superficial staphylococcal lesion frequently precedes the development of bone infection. In the preantibiotic era, *Streptococcus pneumoniae* was a common cause of acute osteomyelitis. *Mycobacterium tuberculosis* and gram-negative organisms are implicated less frequently in this infection.

189. The answer is C. *(Howard, pp 422–423.)* All toxigenic strains of *Corynebacterium diphtheriae* are lysogenic for β-phage carrying the *tox* gene, which codes for the toxin molecule. The expression of this gene is controlled

Bacteriology Answers

by the metabolism of the host bacteria. The greatest amount of toxin is produced by bacteria grown on media containing very low amounts of iron.

190. The answer is E. *(Davis, 4/e. pp 596–597.)* Exotoxin A is produced by more than 90 percent of isolates of *P. aeruginosa*. Synthesis of toxin is induced by iron limitation in the tissues. The mode of action is similar to that of diphtheria toxin. Pyoverdin and pyocyanin are pigments produced by the organism and aid in its laboratory identification.

191. The answer is B. *(Davis, 4/e. pp 548–549.)* Toxic shock syndrome (TSS) is a febrile illness seen predominantly, but not exclusively, in menstruating women. Clinical criteria for TSS include fever greater than 102°F (38.9°C), rash, hypotension, and abnormalities of the mucous membranes and the gastrointestinal, hepatic, muscular, cardiovascular, or central nervous system. Usually three or more systems are involved. Treatment is supportive, including the aggressive use of antistaphylococcal antibiotics. Certain types of tampons may play a role in TSS by trapping O_2 and depleting Mg^{2+}. Most people have protective antibodies to the toxic shock syndrome toxin (TSST-1).

192. The answer is A. *(Davis, 4/e. pp 548–549.)* Toxic shock syndrome (TSS) is caused by a toxin-producing strain of *Staphylococcus aureus* (TSST-1). While there have been reports that *S. epidermidis* produces TSS, they have largely been discounted. Vaginal colonization with *S. aureus* is a necessary adjunct to the disease. *S. aureus* is isolated from the vaginal secretions, conjunctiva, nose, throat, cervix, and feces in 45 to 98 percent of cases. The organism has infrequently been isolated from the blood.

193. The answer is C. *(Davis, 4/e. pp 548–549.)* Epidemiologic investigations suggest strongly that toxic shock syndrome is related to use of tampons, in particular, use of the highly absorbent ones that can be left in for extended periods of time. An increased growth of intravaginal *S. aureus* and enhanced production of TSST-1 have been associated with the prolonged intravaginal use of these hyperabsorbent tampons and with the capacity of the materials used in them to bind magnesium. The most severe cases of TSS have been seen in association with gram-negative infection. TSST-1 may enhance endotoxin activity.

194–197. The answers are: 194-C, 195-D, 196-C, 197-B. *(Lorian, pp 250–251.)* Beysian statistics are often used to determine sensitivity, specificity, and predictive values of new diagnostic tests. A square is set up and the experimental numbers inserted: a = true positive, b = false positive, c = false negative, and d = true negative. The formulas for sensitivity, specificity, and predictive values are also given.

LA Test	Culture	
	POS	NEG
POS	(a) 25	(b) 5
NEG	(c) 5	(d) 95

$$\text{Sensitivity} = \frac{a}{a + c} = \frac{25}{25 + 5} = 85\%$$

$$\text{Specificity} = \frac{d}{d + b} = \frac{95}{95 + 5} = 95\%$$

$$\text{PVP} = \frac{a}{a + b} = \frac{25}{25 + 5} = 85\%$$

$$\text{PVN} = \frac{d}{d + c} = \frac{95}{95 + 5} = 95\%$$

It is necessary to note that the incidence of the disease in the population affects predictive values but not sensitivity or specificity. At a given level of sensitivity and specificity, as the incidence of the disease in the population increases the predictive value of a *positive* (*PVP*) increases and the predictive value of a *negative* (*PVN*) decreases. For this reason, predictive values are difficult to interpret unless true disease incidence is known.

198–200. The answers are: 198-D, 199-B, 200-C. *(Jawetz, 19/e. pp 149–179, 216–217, 327–330.)* These questions demonstrate commonly occurring clinical infectious diseases and microbiologic problems. Enterococci recently have been shown to be resistant to ampicillin and gentamicin. Vancomycin would be the drug of choice. However, laboratory results do not always correlate well with clinical response. The National Committee on Clinical Laboratory Standards recommends testing enterococci only for ampicillin and vancomycin.

Some symptomatic patients may have 10 leukocytes per milliliter of urine but relatively few bacteria. The patient is likely infected and the organisms, particularly if in pure culture, should be further processed.

The patient in the last question of the group probably has actinomycosis. These laboratory data are not uncommon. There is no reason to "work up" all the contaminating bacteria. A fluorescent microscopy test for *Actinomyces israelii* is available. If positive, the FA provides a rapid diagnosis. In any event, it may be impossible to recover *A. israelii* from such a specimen. High-dose penicillin has been used to treat actinomycosis.

201–205. The answers are: 201-C, 202-A, 203-E, 204-A, 205-D. *(Davis, 4/e. pp 206–220.)* There are few bacteria for which antimicrobial susceptibility is highly predictable. However, some agents are the drug of choice because of their relative effectiveness. Among the three antibiotics that have been shown to treat legionellosis effectively (erythromycin, rifampin, and minocycline), erythromycin is clearly superior, even though in vitro studies show the organism to be susceptible to other antibiotics.

Penicillin remains the drug of choice for *Streptococcus pneumoniae* and the group A streptococci, although a few isolates of penicillin-resistant pneumococci have been observed. Resistance among the pneumococci is either chromosomally mediated, in which case the minimal inhibitory concentrations (MICs) are relatively low, or plasmid-mediated, which results in highly resistant bacteria. The same is generally true for *Haemophilus influenzae*. Until the mid-1970s, virtually all isolates of *H. influenzae* were susceptible to ampicillin. There has been a rapidly increasing incidence of ampicillin-resistant isolates, almost 35 to 40 percent in some areas of the U.S. Resistance is ordinarily mediated by β-lactamase, although ampicillin-resistant, β-lactamase-negative isolates have been seen. No resistance to penicillin has been seen in group A streptococci.

Clostridium difficile causes toxin-mediated pseudomembranous enterocolitis as well as antibiotic-associated diarrhea. Pseudomembranous enterocolitis is normally seen during or after administration of antibiotics. One of the few agents effective against *C. difficile* is vancomycin. Alternatively, bacitracin can be used.

Lyme disease, caused by *Borrelia burgdorferi*, has been treated with penicillin and tetracycline. Treatment failures have been observed. Ceftriaxone has become the drug of choice, particularly in the advanced stages of Lyme disease.

206–209. The answers are: 206-D, 207-A, 208-C, 209-B. *(Balows, Clinical Microbiology, 5/e. pp 402–404.)* Some organisms originally thought to be vibrios, such as *Campylobacter jejuni*, have been reclassified. *C. jejuni*, which grows best at 42°C, has its reservoir in birds and mammals and causes gastroenteritis in humans.

Vibrio cholerae causes cholera, which is worldwide in distribution. The three serotypes for cholera are Ogawa (AB), Inaba (AC), and Hikojima (ABC). The isolate of *V. cholerae* is "string-test"-positive.

Vibrio parahaemolyticus is a halophilic marine vibrio that causes gastroenteritis in humans, primarily from ingestion of cooked seafood. It is lactose-negative, sucrose-negative.

Vibrio vulnificus is also halophilic. It has been suggested that these halophilic vibrios do not belong in the genus *Vibrio* but in the genus *Beneckea*. *V. vulnificus* is lactose-positive and produces heat-labile, extracellular toxin. Organisms that, unlike *V. cholerae*, do not agglutinate in 0-1 antiserum were once called *nonagglutinable (NAG)*, or *noncholera (NC)*, vibrios. Such a classification can be confusing because *V. vulnificus*, which is an NCV, nevertheless causes severe cholera-like disease. In addition, *V. vulnificus* can produce wound infections, septicemia, meningitis, pneumonia, and keratitis.

210–213. The answers are: 210-E, 211-C, 212-A, 213-D. *(Jawetz, 19/e. pp 237–244.)* The organisms described in the question all are short, ovoid, gram-negative rods. For the most part, they are nutritionally fastidious and require blood or blood products for growth. These and related organisms are unique among bacteria in that though they have an animal reservoir, they can be transmitted to humans. Humans become infected by a variety of routes, including ingestion of contaminated animal products (*Brucella abortus* in cattle), direct contact with contaminated animal material or with infected animals themselves (*Yersinia enterocolitica* and *Bordetella bronchiseptica* in dogs), and animal bites (*Pasteurella multocida* in many different animals). The laboratory differentiation of these microbes may be difficult and must rely on a number of parameters, including biochemical and serologic reactions, development of specific antibody response in affected persons, and epidemiologic evidence of infection.

214–218. The answers are: 214-C, 215-D, 216-E, 217-B, 218-A. *(Jawetz, 19/e. pp 188–191, 194–199, 230–232, 250–255, 272–277.)* The medium of choice for the isolation of pathogenic neisseriae is Thayer-Martin (TM) agar. TM agar is both a selective and an enriched medium; it contains hemoglobin, the supplement Isovitalex, and the antibiotics vancomycin, colistin, mystatin, and trimethoprim.

Vibrio cholerae as well as other vibrios, including *V. parahaemolyticus* and *V. alginolyticus*, are isolated best on thiosulfate citrate bile salts sucrose medium, although media such as mannitol salt agar also support the growth of vibrios. Maximal growth occurs at a pH of 8.5 to 9.5 and at 37°C incubation.

Löwenstein-Jensen slants or plates, which are composed of a nutrient base and egg yolk, are used routinely for the initial isolation of mycobacteria. Small inocula of *M. tuberculosis* can also be grown in oleic acid albumin media; large inocula can be cultured on simple synthetic media.

Löffler's medium, which is very rich, supports the growth of *Corynebacterium diphtheriae* but suppresses the growth of most other nasopharyngeal microflora. *C. diphtheriae* colonies on this medium appear small, gray, and granular and have uneven edges. *Staphylococcus aureus* grows very well on sheep blood agar, which is made up of a nutrient base and 5 to 8 percent sheep blood; selective and differential media, such as mannitol salt agar, also are available for *S. aureus*.

219–222. The answers are: 219-C, 220-A, 221-E, 222-B. *(Davis, 4/e. pp 728–729.)* An understanding of normal, or indigenous, microflora is essential in order to appreciate the abnormal. Usually, anatomic sites contiguous to mucous membranes are not sterile and have a characteristic normal flora.

The skin flora differs as a function of location. Skin adjacent to mucous membranes may share some of the normal flora of the gastrointestinal system. Overall, the predominant bacteria on the skin surface are *Staphylococcus epidermidis* and *Propionibacterium*, an anaerobic diphtheroid.

The gastrointestinal tract is sterile at birth and soon develops a characteristic flora as a function of diet. In the adult, anaerobes such as *Bacteroides fragilis* and *Bifidobacterium* may outnumber coliforms and enterococci by a ratio of 1000 to 1. The colon contains 10^{11} to 10^{12} bacteria per gram of feces.

The mouth is part of the gastrointestinal tract but its indigenous flora shows some distinct differences. While anaerobes are present in large numbers, particularly in the gingival crevice, the eruption of teeth at 6 to 9 months of age leads to colonization by organisms such as *Streptococcus mutans* and *S. sanguis*, both α-hemolytic streptococci. An edentulous person loses α-hemolytic streptococci as normal flora.

Soon after birth, the vagina becomes colonized by lactobacilli. As the female matures, lactobacilli may still be predominant, but anaerobic cocci, diphtheroids, and anaerobic gram-negative rods also are found as part of the indigenous flora. Changes in the chemical or microbiologic ecology of the vagina can have marked effects on normal flora and may promote infection such as vaginitis or vaginosis.

223–226. The answers are: 223-D, 224-A, 225-C, 226-E. *(Howard, pp 245–261, 405–406.)* *Streptococcus salivarius*, *S. mutans*, *Actinomyces viscosus*, and *A. israelii* are all part of the normal microbiota of the human mouth. Both streptococci are usually α-hemolytic, although nonhemolytic variants may appear, and both are common causes of bacterial endocarditis. *S. mutans* is highly cariogenic (i.e., capable of producing dental caries), in large part because of its unique ability to synthesize a dextran bioadhesive that sticks to teeth. *S. salivarius* settles onto the mucosal epithelial surfaces of the human mouth soon after birth and is often found in the saliva.

Members of the genus *Actinomyces* that are clinically significant can be differentiated by specific fluorescent antibody microscopy as well as a battery of physiologic tests such as those assessing requirements for oxygen. *Actinomyces* organisms are opportunistic members of the normal oral microbiota. Both *A. israelii* and *A. viscosus* are pathogenic and can cause osteomyelitis in the cervicofacial region. Of the two species, *A. israelii*, which is anaerobic, is the more common causative agent of actinomycosis. *A. viscosus*, a facultative anaerobe, appears to be cariogenic.

227–233. The answers are: 227-B, 228-C, 229-D, 230-E, 231-A, 232-F, 233-I. *(Balows, Clinical Microbiology, 5/e. pp 222–258, 277–286, 360–383, 454–456, 471–477. Howard, p 439.)* Diphtheria, a disease caused by *Corynebacterium diphtheriae*, usually begins as a pharyngitis associated with pseudomembrane formation and lymphadenopathy. Growing organisms lysogenic for a prophage produce a potent exotoxin that is absorbed in mucous membranes and causes remote damage to the liver, kidneys, and heart; the polypeptide toxin inhibits protein synthesis of the host cell. Although *C. diphtheriae* may infect the skin, it rarely invades the bloodstream and never actively invades deep tissue. Diphtheria toxin (DT) kills sensitive cells by blocking protein synthesis. DT is converted to an enzyme that inactivates elongation factor 2 (EF-2), which is responsible for the translocation of polypeptidyl-tRNA from the acceptor to the donor site on the eukaryotic ribosome. The reaction is as follows:

$$NAD + EF\text{-}2 = ADP\text{-}ribosyl - EF\text{-}2 + nicotinamide + H^+$$

Bordetella pertussis and *B. parapertussis* are similar and may be isolated together from a clinical specimen. However, *B. parapertussis* does not produce pertussis toxin. Pertussis toxin, like many bacterial toxins, has two subunits: A and B. Subunit A is an active enzyme and B promotes binding of the toxin to host cells.

Francisella tularensis is a short, gram-negative organism that is markedly pleomorphic; it is nonmotile and cannot form spores. It has a rigid growth requirement for cysteine. Human tularemia usually is acquired from direct contact with tissues of infected rabbits but also can be transmitted by the bites of flies and ticks. *F. tularensis* causes a variety of clinical syndromes, including ulceroglandular, oculoglandular, pneumonic, and typhoidal forms of tularemia.

The pathogenesis of infection with *Escherichia coli* is a complex interrelation of many events and properties. *E. coli* may serve as a model for other members of the Enterobacteriaceae. Some strains of *E. coli* are enteroinvasive (EIEC), some enterotoxic (ETEC), some enterohemorrhagic (EHEC), and others enteropathogenic (EPEC). At the present time, there is little clinical significance in routinely discriminating the various types, with the possible exceptions of the ETEC and the 0157/H7 *E. coli* that are hemorrhagic. *E. coli* 0157/H7 secretes a toxin called "*verotoxin.*" The toxin is very active in a Vero cell line. More correctly, the toxin(s) should be called "*Shiga-like.*" Streptococcal infection usually is accompanied by an elevated titer of antibody to some of the enzymes produced by the organism. Among the antigenic substances elaborated by group A β-hemolytic streptococci are erythrogenic toxin, streptodornase (streptococcal DNase), hyaluronidase, and streptolysin O (a hemolysin). Streptolysin S is a nonantigenic hemolysin. Specifically, erythrogenic toxin causes the characteristic rash of scarlet fever.

Many factors play a role in the pathogenesis of *N. meningitidis.* A capsule containing *N*-acetylneuraminic acid is peculiar to *Neisseria* and *E. coli* K1. Fresh isolates carry pili on their surfaces, which function in adhesion. *Neisseria* have a variety of membrane proteins, and their role in pathogenesis can only be speculated upon at this time. The lipopolysaccharide (LPS) of *Neisseria,* more correctly called *lipooligosaccharide (LOS),* is the endotoxic component of the cell.

There are no known toxins, hemolysins, or cell wall constituents known to play a role in the pathogenesis of disease by *Brucella.* Rather, the ability of the organisms to survive within the host phagocyte and to inhibit neutrophil degranulation is a major disease-causing factor. In infectious abortion of cattle caused by *Brucella,* the tropism for placenta and the chorion is a function of the presence of erythritol in allantoic and amniotic fluid.

Physiology

DIRECTIONS: Each question below contains five suggested responses. Select the **one best** response to each question.

234. Penicillinase isolated from *Staphylococcus aureus* inactivates 6-aminopenicillanic acid (shown below) by breaking which of the following numbered bonds?

(A) 1
(B) 2
(C) 3
(D) 4
(E) 5

235. Freeze-etching is a method of preparing cells for electron microscopy. Freeze-etch particles can be located in the

(A) nucleolus
(B) nucleus
(C) cytoplasm
(D) cell wall
(E) cell membrane

236. Gentamicin is an aminoglycoside antibiotic that binds to bacterial ribosomes and inhibits protein synthesis. *Fusobacterium*, an anaerobic gram-negative rod, is resistant to gentamicin, but gentamicin can readily bind to *Fusobacterium* ribosomes, and protein synthesis subsequently ceases. Possible explanations for this phenomenon include that

(A) ribosomes in *Fusobacterium* are protected by a membrane that prevents the binding of gentamicin
(B) gentamicin is inactivated by a specific β-lactamase
(C) gentamicin does not bind to gentamicin-binding proteins that facilitate entry into the cell
(D) gentamicin is taken into the cell by an oxidative active transport system that is nonfunctional under anaerobic conditions
(E) under anaerobic conditions protein synthesis is ribosome-independent

237. In the photomicrograph below, the *Escherichia coli* bacteria shown in cross section are

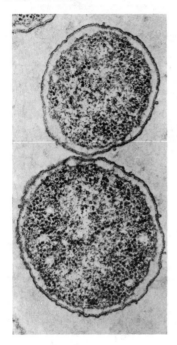

(A) normal in appearance
(B) organizing for mitosis
(C) in a hypotonic environment
(D) partially plasmolyzed
(E) sporulating

238. Bacteria that produce propionic acid by propionic fermentation are

(A) found only in the genus *Propionibacterium*
(B) also producers of ethanol
(C) important in the manufacture of champagne
(D) slow-growing fermenters of pyruvate
(E) obligate aerobic bacteria

239. An aliquot of *Escherichia coli* is treated with ethylenediaminetetraacetic acid (EDTA). The first wash is analyzed and found to contain alkaline phosphatase, DNase, and penicillinase. The anatomic area of the cell affected by the EDTA is most likely to have been the

(A) periplasmic space
(B) mesosomal space
(C) chromosome
(D) plasma membrane
(E) slime layer

240. A genetic probe for the diagnosis of *Mycoplasma pneumoniae* has recently been discovered. It is rapid (2 h), does not require that DNA be digested to produce single strands, shows little cross-reactivity with other bacteria, and is 100 times more sensitive than other probes. This probe is most likely

(A) a DNA probe that binds to double-stranded DNA
(B) a DNA probe that binds to tRNA
(C) a DNA probe that is specific for the genus *Mycoplasma* and binds to cell-wall constituents
(D) an RNA probe that binds to ribosomal RNA
(E) an RNA probe that binds to mRNA

241. The long bacterial structure shown in the electron micrograph below is necessary for

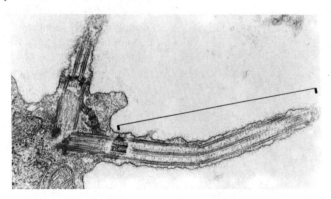

(A) motility
(B) active transport
(C) cellular rigidity
(D) cellular attachment
(E) conjugation

242. Gram-negative and gram-positive bacterial cell walls share which of the following characteristics?

(A) Peptide cross-links between polysaccharides
(B) An equal susceptibility to hydrolysis by lysozyme
(C) Nonfunctional parons
(D) A wide variety of complex lipids
(E) Ketodeoxyoctanate

243. Endospores of *Bacillus subtilis* are characterized by all the following EXCEPT

(A) a lack of metabolic activity
(B) greater resistance than the vegetative cell to drying
(C) multiple covering layers, including a peptidoglycan-containing sporewall cortex
(D) a high calcium content
(E) lack of dipicolinic acid

244. All the statements about the *Escherichia coli* cells shown in the photomicrograph below are true EXCEPT

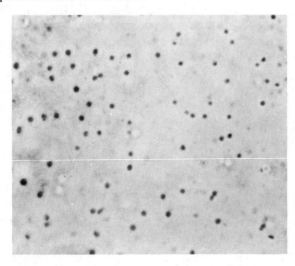

(A) they can result from treatment with penicillin
(B) they can result from treatment with lysozyme
(C) they are osmotically fragile
(D) they are commonly referred to as *endospores*
(E) they have partially formed cell walls

245. After exposure to a mutagenic agent, a strain of *Staphylococcus aureus* is observed to be devoid of division septa. Such a phenomenon most likely would be caused by a deleterious effect on which of the following organelles?

(A) Phagosome
(B) Lysosome
(C) Ribosome
(D) Mesosome
(E) Somatosome

246. A recently hired laboratory technologist forgets the iodine-fixation step while performing a Gram's stain on a strain of *Staphylococcus*. The most likely result is that the organism would

(A) appear pink
(B) appear blue
(C) be colorless
(D) wash off the slide
(E) lyse

247. Analysis of the metabolites produced by an organism's fermentation of glucose reveals small amounts of 6-phosphogluconic acid. This fermentation organism is most likely to be

(A) *Enterobacter*
(B) *Escherichia*
(C) *Leuconostoc*
(D) *Enterococcus faecalis*
(E) *Streptococcus lactis*

248. The formation of adenosine triphosphate (ATP) is essential for the maintenance of life. In mammalian systems, the number of moles of ATP formed per gram atom of oxygen consumed (the P/O ratio) is 3; in bacteria, however, the P/O ratio may be only 1 or 2. The primary reason for the lower P/O ratio in bacteria is

(A) absence of nicotinamide adenine dinucleotide (NAD)
(B) loss of oxidative phosphorylation coupling sites
(C) less dependence on ATP as an energy source
(D) absence of a nonphosphorylative bypass reaction
(E) a less efficient mesosome

249. An aerobic organism is incubated in the presence of acetic acid, which is used as a carbon and energy source. Analysis of the metabolic intermediates reveals, among other substances, succinic acid, acetyl CoA, but no pyruvate. The series of reactions responsible for such a pattern of metabolites is called the

(A) succinate cycle
(B) Krebs cycle
(C) TCA cycle
(D) glyoxylate cycle
(E) Entner-Doudoroff pathway

250. Regulation of branched biosynthetic pathways can be effected by all the following EXCEPT

(A) sequential feedback inhibition
(B) concerted feedback inhibition
(C) cumulative feedback inhibition
(D) enzyme induction
(E) isofunctional enzymes

251. The Mitchell hypothesis for energy conservation in biologic systems

(A) applies only to plants
(B) explains the rapid growth rate of *Escherichia coli*
(C) requires an intact membrane vesicle for proper function
(D) would not apply to bacteria without cytochromes
(E) explains substrate phosphorylation

252. Reversion of *Neisseria gonorrhoeae* from a fimbriated (Fim +) to a nonfimbriated (Fim −) state would result in which one of the following phenomena?

(A) Inability of *N. gonorrhoeae* to colonize the mucosal epithelium
(B) Reversion to Gram positivity
(C) Death of the organism
(D) Loss of serologic specificity
(E) A negative capsule strain

253. An unknown isolate is recognized serologically as *Salmonella enteritidis* serovar newport. A mutant of this organism has lost region 1 (O-specific polysaccharide) of its lipopolysaccharide. This mutant would be identified as

(A) *Salmonella typhi*
(B) *Salmonella newport*
(C) *Salmonella enteritidis*
(D) *Salmonella enteritidis* serovar newport
(E) *Arizona*

254. An *Escherichia coli* auxotrophic mutant for the biosynthesis of methionine is likely to

(A) be temperature-sensitive
(B) grow in a sulfur-containing, methionine-free medium
(C) be resistant to penicillin in a methionine-enriched medium
(D) become actively mitotic in a methionine-enriched medium
(E) utilize arginine instead of methionine for protein biosynthesis

255. Selective inhibition of synthesis of dipicolinic acid (structure shown below) would most likely inhibit the formation of

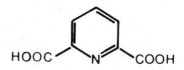

(A) bacterial flagella
(B) bacterial spores
(C) eukaryotic cilia
(D) eukaryotic flagella
(E) fimbriae

256. Protoplasts, spheroplasts, and L forms of bacteria have morphologic and colonial similarities despite the fact that they are taxonomically unrelated. Their morphologic and colonial similarities are related to the

(A) absence of a rigid cell wall
(B) absence of a polysaccharide capsule
(C) presence of a phospholipid outer membrane
(D) presence of endospores
(E) presence of peritrichous flagella

257. Ideally, an antibiotic should focus on a microbial target not found in mammalian cells. By this standard, which of the following antibiotic agents would be expected to be most toxic to humans?

(A) Penicillin
(B) Mitomycin
(C) Cephalosporin
(D) Bacitracin
(E) Vancomycin

258. A freeze-fractured *Escherichia coli* is shown below. The elliptical structure at the left is the

(A) plasma membrane
(B) cell wall
(C) cell capsule
(D) cytoplasm
(E) flagellum

259. In the selection of radiation-induced reversions from a methionine auxotroph, a trace of methionine is required in agar plates of "methionine-free" medium to allow growth of the prototrophic mutant. Requiring an essential factor to produce reversion mutants defines the phenomenon of

(A) induction
(B) mutant resistance
(C) phase variation
(D) phenotypic lag
(E) periodic selection

260. *Escherichia coli* has two major porins located in the outer membrane. The function of porins is

(A) stabilization of the mesosome
(B) metabolism of phosphorylated intermediates
(C) transfer of small molecules through the outer membrane
(D) serologic stabilization of the O antigen
(E) diffusion of safranin from the cell, thereby rendering the cell gram-negative

261. A 21-year-old man was bitten by a tick in Oregon. Two years later, during the course of routine screening for an unknown ailment, a screening Lyme disease test was performed, which was negative. A Western blot strip (IgG) showed the following pattern:

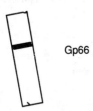 Gp66

Which of the following is the correct interpretation of the test?

(A) The patient has acute Lyme disease
(B) The patient has chronic Lyme disease
(C) The pattern may represent non-specific reactivity
(D) The screening test should be repeated
(E) The patient should be tested for HIV on the basis of the Western blot

Microbiology

DIRECTIONS: Each group of questions below consists of lettered headings followed by a set of numbered items. For each numbered item select the **one** lettered heading with which it is **most** closely associated. Each lettered heading may be used **once, more than once, or not at all.**

Questions 262–265

For each numbered item, choose the lettered growth curve (in an exponentially growing culture) with which it is most likely to be associated. (The arrow in the graph indicates the time at which the drugs were added.)

Questions 266–268

For each description below, select the process with which it is most likely to be associated.

(A) Conjugation
(B) Recombination
(C) Competence
(D) Transformation
(E) Transduction

266. Uptake by a recipient cell of soluble DNA released from a donor cell

267. Transfer of a donor chromosome fragment by a temperate bacterial virus

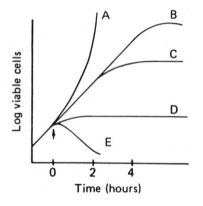

268. Direct transfer of a plasmid between two bacteria

262. Chloramphenicol

263. Penicillin

264. Sulfonamide

265. Control (without antibiotic)

Questions 269–272

A 7% sodium dodecyl sulfate polyacrylamide gel electrophoretogram of *Escherichia coli* cell walls is shown below. For each numbered item, choose the lettered band on the electrophoretogram with which it is most likely to be associated.

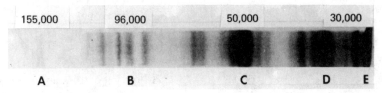

269. Lactose permease

270. β and β' RNA polymerase

271. Flagellin

272. Major cell-wall polypeptide

Questions 273–277

For each bacterium listed below, select the metabolic reaction with which it is most characteristically associated.

(A) Mixed acid fermentation
(B) Homolactic fermentation
(C) Yeast alcohol fermentation
(D) Production of 2,3-butanediol
(E) Production of propionic acid

273. *Klebsiella*

274. *Lactobacillus*

275. *Arachnia*

276. *Escherichia coli*

277. *Enterobacter*

Questions 278–280

For each numbered structure listed, match the appropriate lettered structure in the freeze-fractured *Escherichia coli* cell shown below.

(A) Structure A
(B) Structure B
(C) Structure C
(D) Structure D
(E) None of the above

278. Plasma membrane

279. Eutectic layer

280. Cell wall (lipoid layer)

Questions 281–285

Match the descriptions below with the appropriate antibiotic.

(A) Piperacillin
(B) Cefoperazone
(C) Ceftriaxone
(D) Ciprofloxacin
(E) Imipenem

281. A broad-spectrum antibiotic in the general class of thienamycins

282. Third-generation cephalosporin with good activity against *Borrelia burgdorferi*

283. Broad-spectrum penicillin with antipseudomonas activity

284. Third-generation cephalosporin with primary activity against *Pseudomonas aeruginosa*

285. Quinolone antibiotic with broad gram-negative and gram-positive activity

Questions 286–291

Match the numbered bacteria and bacterial components with the lettered chemical constituent that is found in them.

(A) Phospholipid
(B) Ribitol teichoic acid
(C) Glycolipids (waxes)
(D) Ketodeoxyoctanate
(E) Peptidoglycan
(F) Outer membrane proteins
(G) Gamma glutamyl polypeptide
(H) Sialic acid polymers
(I) Hyaluronic acid

286. *Neisseria meningitidis*, group B

287. Group A streptococci

288. Bacterial lipopolysaccharide (LPS)

289. *Mycobacterium* cell walls

290. *Bacillus anthracis* capsules

291. All bacteria

Questions 292–296

Match the mechanism of action with the correct antibiotic.

(A) Penicillin
(B) Amdinocillin
(C) Amphotericin
(D) Chloramphenicol
(E) Trimethoprim

292. Inhibits dihydrofolate reductase

293. Binds to penicillin-binding protein-2 (PBP-2)

294. Inhibits the final peptide bond between D-alanine and glycine

295. Binds sterols and alters membrane permeability

296. Attaches to 50S ribosome and inhibits peptidyl transferase

Physiology

Answers

234. The answer is D. *(Jawetz, 19/e. pp 161–163.)* The structural integrity of the β-lactam ring in penicillins is essential for their antimicrobial activity. Many resistant strains of staphylococci produce an enzyme, penicillinase, that cleaves the β-lactam ring at the carbon-nitrogen bond. Other organisms, including certain coliform bacteria, produce an amidase enzyme that inactivates penicillin by disrupting the bond between the radical and nitrogen in the free amino group (1 in the diagram).

235. The answer is E. *(Davis, 4/e. pp 30–43.)* As a result of freeze-etching, the bacterial cell-membrane bilayer is split, revealing freeze-etch particles. The particles are thought to represent globular proteins within the hydrophobic region of the membrane. Freeze-etching eliminates many of the alterations induced by other methods of fixation.

236. The answer is D. *(Jawetz, 19/e. p 173.)* There is evidence that gentamicin has binding sites on both 30S and 50S ribosomes, which results in reduced activity of the initiation complex, interference with attachment of tRNA, and distortion of codons. Active transport of gentamicin, however, is energy-dependent and an oxidative process. Aminoglycosides, then, appear not to be transported across the cell membrane in anaerobic bacteria. All anaerobes are resistant to aminoglycosides.

237. The answer is D. *(Jawetz, 19/e. pp 7–31.)* If a bacterial cell, especially a gram-negative cell, is exposed to a hypertonic solution, the cell membrane and contents contract and shrink away from the cell wall. This phenomenon is called *plasmolysis*. The presence of a rigid cell wall outside the cytoplasmic membrane is a distinctive feature of plasmolysis.

238. The answer is D. *(Davis, 4/e. pp 70–75.)* *Propionibacterium acnes* and *Arachnia propionica* are capable of fermenting pyruvate to propionic acid and carbon dioxide and generating ATP in the process. This pathway extracts more energy from substrate than is extracted by alcoholic fermentation. Ethanol is not a by-product of propionic acid fermentation. Propionic fermentation is important in the commercial manufacture of Swiss cheese; carbon dioxide is responsible for the holes, and propionic acid for the odor and flavor.

Physiology Answers

239. The answer is A. *(Jawetz, 19/e. pp 7–31.)* The periplasm is the space between a bacterium's outer membrane and plasma membrane. The periplasmic space in *Escherichia coli* has been shown to contain a number of proteins, sugars, amino acids, and inorganic ions. Ethylenediaminetetraacetic acid (EDTA) is a chelating agent that disrupts the cell walls of gram-negative bacteria.

240. The answer is D. *(Jawetz, 19/e. p 596.)* A probe with the characteristics stated in the question is most certainly an RNA probe. Ribosomal RNA is single-stranded; hence no digestion of double-stranded DNA is required. There is much more ribosomal RNA than DNA, mRNA, or tRNA. Cross-reaction with other bacteria is not surprising, as ribosomal RNA may have some similar sequences. Most probes marketed are DNA probes. Present tests are moderately rapid (2 to 3 h). Probe technology is expected to produce much more rapid (less than 1 h) tests in the near future.

241. The answer is A. *(Jawetz, 19/e. pp 23–26.)* Flagella are organelles of motility. They are long, filamentous structures originating from a spherical basal body. A flagellum is composed of three parts: the filament, the hook, and the basal body.

242. The answer is A. *(Davis, 4/e. pp 30–43.)* A peptidoglycan framework is the basis of both gram-positive and gram-negative bacterial cell walls. This complex network, which imparts rigidity to the cell, is the site of action for lysozyme hydrolysis (gram-negative cell walls, it should be noted, are less susceptible to lysozyme than are gram-positive cell walls). The cell walls of gram-negative organisms are rich in lipopolysaccharides and other complex lipids; gram-positive cell walls, on the other hand, are lipid-poor.

243. The answer is E. *(Davis, 4/e. pp 45–49.)* All spores, including those of *Bacillus subtilis*, exhibit all the characteristics listed in the question, except lack of dipicolinic acid. In fact, dipicolinic acid is a unique characteristic of bacterial spores. The ability to form spores is a characteristic of three groups of gram-positive organisms: clostridia, bacilli, and sporosarcinae. The function of calcium in spores may be to contract the loose polyanionic cortical peptidoglycan and thus expel water and contribute to structural strength. Sporulation is initiated when environmental conditions become unfavorable. Depletion of nitrogen and carbon is thought to play a key role.

244. The answer is D. *(Baron, 3/e. pp 48–49.)* The organisms illustrated in the question are spheroplasts of *Escherichia coli*. Lysozyme cleans the β-1-4-glycosidic bond between *N*-acetylmuramic acid and *N*-acetylglucosamine.

Spheroplasts are bacteria with cell walls that have been partially removed by the action of lysozyme or penicillin. Ordinarily, with disintegration of the walls, the cells undergo lysis; however, in a hypertonic medium, the cells persist and assume a spherical configuration. Endospores are formed by gram-positive bacteria in the genera *Bacillus* and *Clostridium*. It has also been shown that for *E. coli* and other gram-negative rods, exposure to minimal concentrations of antibiotics (1/4 × MIC) does not rupture the cell wall but promotes elongation of the cell by inhibiting the division cycle.

245. The answer is D. *(Jawetz, 19/e. pp 7–31.)* Mesosomes are specialized bacterial organelles often associated with division septa. Mesosomes are formed from invaginations of the cell membrane and aid in the development of cross-walls during cell division. They are also the site of attachment of the bacterial chromosome.

246. The answer is A. *(Jawetz, 19/e. p 29.)* Gram's staining method involves the application of a basic dye, crystal violet, followed by iodine for fixation. The preparation then is treated with ethanol or acetone, which decolorizes gram-negative bacteria. Finally, a red counterstain, safranin, is applied to restain the decolorized organisms. Failure to include the iodine step would prevent the formation of the crystal violet–iodine complex; consequently, the organism would stain gram-negative (pink).

247. The answer is C. *(Jawetz, 19/e. pp 59–66.)* 6-Phosphogluconic acid is a characteristic metabolic intermediate in the pentose-phosphate metabolic pathway. This pathway is used by heterolactic fermenters such as *Leuconostoc*, the organism responsible for the fermentation of cabbage in the production of sauerkraut. *Leuconostoc* is a gram-positive bacterium with a dextran capsule.

248. The answer is B. *(Baron, 3/e. pp 80–81.)* ATP is believed to be generated at three reaction points in the electron transport chain: the reductions of flavoprotein, cytochrome *b*, and cytochrome *c*. This phenomenon, demonstrated in experiments with mammalian mitochondria, can be expressed in terms of the relationship between the moles of ATP generated for each atom of oxygen consumed—the P/O ratio. In mammalian cells, the P/O ratio is 3; that is, there are three segments in the electron transfer chain in which there is a relatively large free energy drop. In bacteria, however, there appears to be only one or two of these segments. Loss of these phosphorylation sites as well as reactions that bypass these sites of ATP synthesis accounts for the lower P/O ratio in bacteria. Some bacteria, such as *Mycobacterium phlei*, have P/O ratios of 3.

Physiology Answers

249. The answer is D. *(Jawetz, 19/e. pp 59–64.)* In the glyoxylate cycle, acetic acid is oxidized without the formation of pyruvic acid. The net result is the conversion of two acetyl residues to succinic acid. Several enzymatic reactions are common to both the glyoxylate cycle and the tricarboxylic acid (TCA) cycle.

250. The answer is D. *(Jawetz, 19/e. pp 70–71.)* A branched biosynthetic pathway is one in which several substances are formed from a common starting point (e.g., aspartic acid is converted to both lysine and methionine). Regulation of these pathways is by a number of processes. Isofunctional enzymes are different enzymes with the same catalytic activity. One isofunctional enzyme may be inhibited by one amino acid end product, whereas a second enzyme is similarly controlled by a different amino acid. Branched pathways also display three types of feedback inhibition: sequential, concerted, and cumulative. Sequential inhibition occurs when two end products inhibit initial steps in their own branches. In concerted and cumulative feedback inhibition, the enzyme catalyzing the first steps of a branched pathway possesses multiple effector sites, each of which binds a different end product.

251. The answer is C. *(Baron, 3/e. p 84.)* The Mitchell hypothesis is complex but it helps explain energy conservation in biologic systems. Energy is generated by the intact membrane-bound electron transport system. Essentially, a proton pump couples energy production to ATP synthesis at the membrane. A proton gradient, even in bacteria that lack a cytochrome-dependent electron transport system, drives all bioenergetic reactions.

252. The answer is A. *(Davis, 4/e. pp 24–25.)* Bacteria may shift rapidly between the fimbriated (Fim +) and the nonfimbriated (Fim −) states. Fimbriae function as adhesions to specific surfaces and consequently play a major role in pathogenesis. Lack of fimbriae prevents colonization of the mucosal surface by the bacterium.

253. The answer is C. *(Jawetz, 19/e. pp 20–21.)* Region 1 (the O-antigenic side chain of lipopolysaccharide) is responsible for the many serotypes of *Salmonella*. A mutant of *Salmonella* deficient in region 1 is not identified as a "newport," at least by virtue of its somatic antigen; biochemical identification of this mutant would be *S. enteritidis*. Loss of region 1 does not affect genus and species classification of *Salmonella*. Recently, however, it has been recommended that *Salmonella* be referred to by genus and serovar, that is, *Salmonella newport* or *Salmonella* serovar newport.

254. The answer is D. *(Davis, 4/e. pp 91–98.)* A methionine auxotrophic mutant cannot synthesize methionine and therefore is unable to grow in a methionine-free medium. Auxotrophs must be grown in media enriched with the essential components (e.g., methionine) that they are unable to produce. Although the growth of mutants may be temperature-sensitive in enriched media, this characteristic is not true of all methionine-requiring auxotrophs. Because penicillin attacks actively multiplying cells, auxotrophs grown in enriched media are susceptible to the drug.

255. The answer is B. *(Jawetz, 19/e. pp 26–27.)* Dipicolinic acid, formed in the synthesis of diaminopimelate (DAP), is a prominent component of bacterial spores but is not found in vegetative cells or eukaryotic appendages or fimbrial structures. The calcium salt of dipicolinic acid apparently plays an important role in stabilizing spore proteins, but its mechanism of action is unknown. Dipicolinic acid synthetase is an enzyme unique to bacterial spores.

256. The answer is A. *(Jawetz, 19/e. pp 22–23.)* Protoplasts are gram-positive bacteria and spheroplasts are gram-negative bacteria that have had their rigid cell walls digested by lysozymes. They must be maintained in a hypertonic medium to prevent lysis. L forms, which are bacteria with defective or absent walls, arise spontaneously under favorable conditions (such as high salt concentrations) or when wall synthesis is impaired by penicillin. Their role in human disease is unclear.

257. The answer is B. *(Davis, 4/e. pp 180–181, 217–218.)* Ideally, antibiotics should attack a microbial structure or function not found in human cells. Except for mitomycin, all the antibiotics listed in the question interfere with cell-wall synthesis in bacteria. Mitomycin inhibits DNA synthesis in both mammalian and microbial systems; viral DNA synthesis, however, is relatively resistant to mitomycin.

258. The answer is A. *(Davis, 4/e. pp 30–43.)* Freeze-etching involves the freezing of cells at very low temperatures in a block of ice. The ice block is split with a knife, and ice crystals are sublimed—etched—from one of the newly exposed faces. The line of fracture often passes through a natural cleavage plane—in the illustration in the question, for example, the inner and outer faces of the cell membrane of an *Escherichia coli*. Freeze-etching does not produce the troublesome artifacts introduced during the fixation and drying of specimens.

259. The answer is D. *(Davis, 4/e. pp 126–127.)* An auxotrophic organism is unable to survive without provision of a specific nutrient not required by the organism's parent (prototrophic) strain. If a prototrophic reversion is created by treating the auxotrophs with a mutagen (e.g., radiation), supply of the essential nutrient must continue for a brief period of time. During this brief phenotypic lag, the newly mutated auxotroph grows and divides, allowing segregation of the mutant recessive allele and cytoplasmic expression of the new gene product.

260. The answer is C. *(Davis, 4/e. pp 41–42.)* *E. coli* has two major porins, Omp C and Omp E (Omp = outer-membrane protein). A porin is a protein trimer with each subunit containing a pore with a diameter of 1 nm. Porins function in outer-membrane (OM) permeability. While porins are known to permit the transfer of small molecules across the OM, specific porins may also influence the diffusion of layer molecules. Depending on charge, porins may also repel certain molecules, such as bile salts found in the intestinal environment.

261. The answer is C. *(Davis, 4/e. p 682. Jawetz, 19/e. p 272.)* The serologic diagnosis of Lyme disease is fraught with difficulty. Enzyme immunoassay (EIA) may be insensitive in the early stages of disease and may lack specificity in advanced stages. Western blot analysis of antibody is the confirmatory test for Lyme disease, but it too is not 100 percent sensitive and specific. The Western blot test detects antibodies to proteins and glycoproteins of *B. burgdorferi*. Not all of these proteins are specific for the organism. For example, antibodies to Gp66 may reflect a cross-reaction as many gram-negative bacteria have similar glycoproteins. For this reason, a Western blot showing only antibodies to Gp66 is thought to be a nonspecific immune response.

262–265. The answers are: 262-D, 263-E, 264-C, 265-B. *(Davis, 4/e. pp 201–228.)* Penicillin causes lysis of growing bacterial cells. Its antimicrobial effect stems from impairment of cell-wall synthesis. Because penicillin is bactericidal, the number of viable cells should fall immediately after introduction of the drug into the medium.

Both chloramphenicol and sulfonamides are bacteriostatic—that is, they retard cell growth without causing cell death. Chloramphenicol causes an immediate, reversible, bacteriostatic inhibition of protein synthesis. Sulfonamides, on the other hand, compete with para-aminobenzoic acid in the synthesis of folate; intracellular stores of folate are depleted gradually as the cells continue to grow.

The number of viable cells in a culture eventually will level off even if no antibiotic is added to the environment. A key factor in this phenomenon is the limited availability of substrate.

266–268. The answers are: 266-D, 267-E, 268-A. *(Jawetz, 19/e. pp 82–94.)* Transformation, transduction, and conjugation are critical processes in which DNA is transferred from one bacterium to another. Transformation, the passage of high-molecular-weight DNA from one bacterium to another, was first observed in pneumococci. Later studies have shown that, at least in *Streptococcus pneumoniae,* double-stranded DNA is "nicked" by a membrane-bound endonuclease, initiating DNA entry into the host cell. One of the nicked DNA strands is digested, and the other is integrated into the host genome.

In conjugation, too, DNA is passed from one bacterium to another. However, instead of the transfer of soluble DNA, a small loop of DNA, called a plasmid, is passed between cells. Examples of plasmids are the sex factors and the resistance (R) factors.

Transduction is a process in which a fragment of donor chromosome is carried to a recipient cell by a temperate virus (bacteriophage). Transduction, which can affect many bacteria, can be "generalized" or "restricted." In generalized transduction, the phage virus can carry any segment of the donor chromosome; in restricted transduction, the phage carries only those chromosomal segments immediately adjacent to the site of prophage attachment.

269–272. The answers are: 269-D, 270-A, 271-C, 272-C. *(Davis, 4/e. p 138.)* Gel electrophoresis provides a rapid method for identifying bacterial proteins and estimating molecular weights. A gel can be made of a number of substances, including starch, agar, and polyacrylamide. Starch gel has high separating power because the fine gel pores act as a molecular sieve. Agar gel is easier to prepare than starch; separation of proteins is accomplished in 30 to 60 min. Polyacrylamide gel also separates on the principle of the molecular sieve. It is chemically inert and electrically neutral. The biggest disadvantage of polyacrylamide is that its separating powers are so good that protein patterns, or patterns of other heterogeneous substances, may be too complex to interpret. In the electrophoretogram presented in the question, band A represents RNA polymerases (molecular weight 155,000), band C represents both flagellin and the major cell-wall protein (50,000), and band D represents lactose permease (30,000). Band E is the dye front.

273–277. The answers are: 273-D, 274-B, 275-E, 276-A, 277-D. *(Davis, 4/e. pp 69–78.)* Mixed acid fermentation is characteristic of *Escherichia coli.* In this process, substrate is fermented either to lactate or, by the splitting of pyruvate, to formate. *E. coli* also can split formic acid into hydrogen and carbon dioxide in a reaction catalyzed by the enzyme formic hydrogenlyase.

Lactobacillus and most streptococci convert 1 mole of glucose to 2 moles of lactic acid in a process known as *homolactic fermentation.* (In heterolactic fermentation, only 1 mole of lactate is produced for every mole of glucose.) The souring of milk is a by-product of lactic fermentation.

The production of 2,3-butanediol is characteristic of most *Enterobacter*, *Klebsiella*, and *Serratia* species. Synthesis of this alcohol is the basis for the positive Voges-Proskauer reaction in these organisms.

Propionic acid is produced by some gram-positive, non-spore-forming rods, such as *Propionibacterium* and *Arachnia*. Propionic acid contributes to the taste and smell of Swiss cheese.

278–280. The answers are: 278-A, 279-C, 280-D. *(Davis, 4/e. pp 30–43.)* Freeze fracture is a process in which cells frozen at $-150°C$ are cleaved with a knife. Ice is sublimed from the cleaved surface, and underlying structures are laid bare. The fracture lines in the ice often pass through cells along natural lines of cleavage and reveal internal surfaces through shadowing on microscopy. Natural bacterial cell planes of cleavage occur between the peptidoglycan layer and the plasma membrane and between the inner and outer faces of the membrane. In the freeze-fracture photograph presented in the question, the concave fractures from the inside of the envelope out include the plasma membrane (A), peptidoglycan layer (B), and the lipopolysaccharide layer (D). Structure C is the eutectic layer.

281–285. The answers are: 281-E, 282-C, 283-A, 284-B, 285-D. *(Howard, pp 121–134.)* Many new antibiotics have become available during the past few years. Although expensive, these antibiotics generally have a broader spectrum of effectiveness than the ones they are intended to replace. Resistance to these newly introduced agents may be a problem that will minimize their effect on the treatment of infectious disease. While most are labeled "broad spectrum," each appears to be characteristically more effective against some organisms than others.

Ceftriaxone is a new-generation cephalosporin. It is administered once a day either intravenously or intramuscularly. While ceftriaxone is used against a wide variety of gram-negative rods, it has found special use in the treatment of Lyme disease. It is now claimed to be the most effective antibiotic for borreliosis.

Imipenem, a representative of the thienamycin class of antibiotics, is biochemically a carbapenem. Imipenem has a wide spectrum of activity against both gram-negative and gram-positive bacteria. Its principal lethal target in the cell is penicillin-binding protein-2. Imipenem is resistant to hydrolysis by most of the known β-lactamases.

Piperacillin is a broad-spectrum, synthetically substituted penicillin similar to carbenicillin and ticarcillin. It has a wider gram-negative spectrum, however, than does carbenicillin or ticarcillin. Piperacillin is effective against most strains of *Pseudomonas aeruginosa*. While it can be used as a single agent, it can also be combined with aminoglycosides such as tobramycin.

Piperacillin is not absorbed orally, is excreted by the kidney, and is 50 percent protein-bound.

Cefoperazone is a third-generation cephalosporin. Although widely treated as "broad spectrum," it has (like most other third-generation cephalosporins) poor activity against gram-positive bacteria. Cefoperazone appears to be an excellent antibiotic for *Pseudomonas aeruginosa*. It is not absorbed orally, is excreted by the kidney, and is 90 percent protein-bound.

Ciprofloxacin is one representative of a new class of antibiotics known as *organic acids*, or *quinolones*. Norfloxacin is a similar antibiotic. Both are related structurally to nalidixic acid, an antimicrobial agent used primarily for urinary tract infection. Ciprofloxacin shows a wide spectrum of activity against virtually all gram-negative bacteria. Although it inactivates gram-positive bacteria, it is least effective against the enterococci. The early trials of the antibiotic were limited to urinary tract infections. Recently, however, trials have been initiated to evaluate its role as a systemic antibiotic. Because of their wide range of activity and minimum toxicity, the quinolones are an exciting group of antibiotics.

286–291. The answers are: 286-H, 287-I, 288-D, 289-C, 290-G, 291-E. *(Baron, 3/e. pp 41–54.)* Bacteria have a variety of components; some are unique to certain genera and species, others are characteristic of all bacteria. All bacteria have peptidoglycan in their cell walls, although the peptidoglycan layer is much thinner in gram-negative than gram-positive bacteria. In gram-positive bacteria, teichoic acids, polysaccharides, and peptidoglycolipids are covalently attached to the peptidoglycans. While *Mycobacterium* also has peptidoglycan, up to 40 percent of the cell wall may be a waxy glycolipid that is responsible for the "acid fastness" of *Mycobacterium* and *Nocardia,* an aerobic actinomycete. Bacterial lipopolysaccharide (LPS), also known as *endotoxin,* is found in only gram-negative bacteria. Not only is it a toxic macromolecule, but it also imparts serologic specificity to some gram-negative bacteria such as *Salmonella* and *E. coli.*

Capsules are found in both gram-positive and gram-negative bacteria. With the exception of those found in *Bacteroides fragilis,* capsules are not in and of themselves toxic but rather are antiphagocytic and are immunologic (or serologic) determinants. Some examples of capsular components are the following:

1. Sialic acid polymers are found in group B *Neisseria meningitidis*. This identical polymer is also found in *E. coli* K1.

2. Group A streptococci in the early stages of growth have hyaluronic acid capsules. The capsule, however, is rapidly destroyed by the organism's own hyaluronidase.

3. *Bacillus anthracis,* the causative agent of anthrax, is the only bacterium to possess a polypeptide capsule that is a polymer of glutamic acid.

Physiology Answers

292–296. The answers are: 292-E, 293-B, 294-A, 295-C, 296-D. *(Jawetz, 19/e. pp 149–152.)* The antibiotics in these questions have significantly different modes of action. Recent evidence suggests that while penicillin inhibits the final cross-linking of the cell wall, it also binds to penicillin-binding proteins and inhibits certain key enzymes involved in cell-wall synthesis. The mechanism is complex. Amdinocillin, although classified as a penicillin, selectively binds to penicillin-binding protein-2 (PBP-2). Binding to PBP-2 results in aberrant cell-wall elongation and spherical forms, seen when *E. coli*, for example, is exposed to mecillinam.

Because amphotericin binds to sterols (such as cholesterol) in the cell membrane, its range of activity is predictable; that is, it is effective against microorganisms that contain sterol in the cell membrane (such as molds, yeasts, and certain amebae). These polyene antibiotics cause reorientation of sterols in the membrane, and membrane structure is altered to the extent that permeability is affected. If sterol synthesis is blocked in fungi, then amphotericin is not effective. This occurs when fungi are exposed to miconazole, another antifungal antibiotic.

Chloramphenicol is a bacteriostatic antibiotic. Its action does not kill the cell but only inhibits it. If chloramphenicol is removed from the culture, then protein synthesis is reinitiated. Bacterial ribosomes are spherical particles with a molecular weight of 3×10^6. Protein synthesis takes place on the ribosome by a complex process involving various ribosomal subunits, tRNA, and nRNA. Chloramphenicol, in contrast to the aminoglycosides and tetracycline, attaches to the 50S ribosome subunit. The enzyme peptidyl transferase, found in the 50S subunit, is inhibited. Removal of the inhibition—in this case chloramphenicol—results in full activity of the enzyme.

Trimethoprim (TMP), a diaminopyrimidine, is a folic acid antagonist. Although TMP is commonly used in combination with sulfa drugs, its mode of action is distinct. TMP is structurally similar to the pteridine portion of dihydrofolate and prevents the conversion of folic acid to tetrahydrofolic acid by inhibition of dihydrofolate reductase. Fortunately, this enzyme in humans is relatively insensitive to TMP.

Rickettsiae, Chlamydiae, and Mycoplasmas

DIRECTIONS: Each question below contains five suggested responses. Select the **one best** response to each question.

297. Which of the following is a newly identified species of *Chlamydia* that causes acute respiratory disease in adults and is not associated with avian sources?
(A) *C. pneumoniae*
(B) *C. trachomatis*
(C) *C. psittaci*
(D) *C. lymphovenereum*
(E) *C. hominis*

298. Mycoplasmas differ from chlamydiae in that they are
(A) susceptible to penicillin
(B) able to grow on artificial cell-free media
(C) able to cause urinary tract infection
(D) able to stain well with Gram's stain
(E) able to cause disease in humans

299. Mycoplasmas are bacterial cells that
(A) fail to reproduce on artificial media
(B) have a rigid cell wall
(C) are resistant to penicillin
(D) stain well with Gram's stain
(E) are energy parasites

300. Brill-Zinsser disease is caused by
(A) *Rickettsia rickettsii*
(B) *R. prowazekii*
(C) *R. typhi*
(D) *R. conorii*
(E) *R. tsutsugamushi*

301. A man with chills, fever, and headache is thought to have "atypical" pneumonia. History reveals that he raises chickens and that approximately 2 weeks ago he lost a large number of them to an undiagnosed disease. The most likely diagnosis of this man's condition is
(A) anthrax
(B) Q fever
(C) relapsing fever
(D) leptospirosis
(E) ornithosis (psittacosis)

302. An ill patient denied being bitten by insects. However, he had spent some time in a milking barn and indicated that it was dusty. Of the following rickettsial diseases, which one has he most likely contracted?

(A) Scrub typhus
(B) Rickettsialpox
(C) Brill-Zinsser disease
(D) Q fever
(E) Rocky Mountain spotted fever

303. A 36-year-old man presented at his physician's office complaining of fever and headache. On examination he had leukopenia and increased liver enzymes and inclusion bodies were seen in his monocytes. History revealed that he was an outdoorsman and remembered removing a tick from his leg. Which of the following diseases should be *excluded* from the differential diagnosis?

(A) Lyme disease
(B) Erhlichiosis
(C) Rocky Mountain spotted fever
(D) Q fever
(E) Tularemia (*Francisella tularensis*)

304. Typhus, spotted fever, and scrub typhus share all the following manifestations of disease EXCEPT

(A) short incubation period (<48 h)
(B) fever
(C) rash
(D) rickettsemia
(E) focal vasculitis

305. Which of the following mycoplasmas has been implicated as a cause of nongonococcal urethritis (NGU)?

(A) *Mycoplasma hominis*
(B) *M. pneumoniae*
(C) *M. fermentans*
(D) *M. mycoides*
(E) *Ureaplasma urealyticum*

306. The mycoplasmas are distinguished from true bacteria by their lack of

(A) a cell wall
(B) lipopolysaccharide
(C) flagella
(D) ATP synthesis
(E) a capsule

307. Lymphogranuloma venereum (LGV) is a venereal disease caused by serotype L_1, L_2, or L_3 of *Chlamydia trachomatis*. The differential diagnosis should include which of the following?

(A) Psittacosis
(B) Chancroid
(C) Shingles
(D) Babesiosis
(E) Mononucleosis

308. Epidemiologically, *Chlamydia trachomatis* infections can be divided into all the following categories EXCEPT

(A) classic trachoma infection
(B) sexually transmitted genital infection in adults
(C) perinatal conjunctivitis
(D) middle-ear infection in young children
(E) infant pneumonia

309. All the following statements about lymphogranuloma venereum (LGV) are true EXCEPT
(A) the causative agent is *Chlamydia trachomatis*
(B) in the U.S., it is more common among women
(C) it is most common in tropical regions
(D) tetracycline is effective in early treatment
(E) late stages of the disease often require surgery

310. An inhibitor of ATP synthesis would be expected to retard most severely the penetration of the host cell by which of the following organisms?
(A) *Chlamydia psittaci*
(B) *Chlamydia trachomatis*
(C) *Ureaplasma urealyticum*
(D) *Rickettsia rickettsii*
(E) *Mycoplasma pneumoniae*

311. *Chlamydia trachomatis* can be distinguished from *Chlamydia psittaci* by which of the following criteria?
(A) *C. trachomatis* is sensitive to sulfonamides
(B) *C. trachomatis* has a different lipopolysaccharide antigen
(C) *C. trachomatis* can be stained with Giemsa
(D) *C. psittaci* is an obligate prokaryotic parasite
(E) *C. psittaci* forms inclusions that contain glycogen

312. All the following statements regarding trachoma are true EXCEPT
(A) it is caused by *Chlamydia trachomatis*
(B) it is best treated with systemic sulfonamides and ophthalmic tetracycline
(C) it affects 400 million people worldwide
(D) it is a form of chronic keratoconjunctivitis
(E) it can occur in animals other than humans

313. Chlamydiae have an unusual three-stage cycle of development. The correct sequence of these events is
(A) penetration of the host cell, synthesis of elementary body progeny, development of an initial body
(B) penetration of the host cell, development of an initial body, synthesis of elementary body progeny
(C) development of an initial body, synthesis of elementary body progeny, penetration of the host cell
(D) synthesis of elementary body progeny, development of an initial body, penetration of the host cell
(E) synthesis of elementary body progeny, penetration of the host cell, development of an initial body

314. Rickettsiae are gram-negative bacteria that cause a wide range of disease. The agent of Rocky Mountain spotted fever is best characterized by the statement that it

(A) grows on 7% sheep blood agar
(B) has an "atypical" gram-negative cell wall
(C) is energy-deficient and cannot phosphorylate glucose
(D) is normal flora of the mosquito gut
(E) is susceptible to penicillin

315. Chlamydiae are small gram-negative rods once thought to be viruses. All the following characteristics of chlamydiae can distinguish them from viruses EXCEPT

(A) independent synthesis of proteins
(B) susceptibility to antibiotics
(C) reproduction by fission
(D) synthesis of ATP
(E) ready visualization with light microscope

316. Q fever is different from all other rickettsial infections in that

(A) it is associated with a skin rash
(B) it is caused by an organism stable outside the host cell
(C) patients' sera contain Weil-Felix antibodies
(D) the causative agent is transmitted by rodents
(E) it can be treated with antibiotics

Microbiology

DIRECTIONS: Each group of questions below consists of lettered headings followed by a set of numbered items. For each numbered item select the **one** lettered heading with which it is **most** closely associated. Each lettered heading may be used **once, more than once, or not at all.**

Questions 317–320

For each clinical or laboratory characteristic, select the species of *Mycoplasma* or *Ureaplasma* with which it is most closely associated.

(A) *Mycoplasma hominis*
(B) *M. orale*
(C) *M. pneumoniae*
(D) *M. fermentans*
(E) *Ureaplasma urealyticum*

317. Causes primary atypical pneumonia in humans

318. Is associated with nongonococcal urethritis in humans

319. Normally inhabits the healthy human oral cavity

320. Normally inhabits the female genital tract but may cause acute respiratory illness

Questions 321–325

For each description below, select the organism with which it is most closely associated.

(A) *Rochalimaea quintana*
(B) *Coxiella burnetii*
(C) *Rickettsia rickettsii*
(D) *Chlamydia trachomatis*
(E) *Chlamydia psittaci*

321. Causative agent of Rocky Mountain spotted fever

322. Transmission by human body louse

323. Causative agent of Q fever

324. Causative agent of a subacute-to-chronic endocarditis

325. Causative agent of lymphogranuloma venereum

Rickettsiae, Chlamydiae, and Mycoplasmas
Answers

297. The answer is A. *(Davis, 4/e. p 706.)* A distinct group of chlamydiae, first designated "TWAR," has been given the name *C. pneumoniae*. The strain was first isolated in Taiwan and causes acute respiratory disease. *C. trachomatis* causes genital and eye infections and special serovars cause lymphogranuloma venereum. *C. psittaci* also causes a respiratory syndrome but is associated with avian contact. *Hominis* is a species of *Mycoplasma*.

298. The answer is B. *(Jawetz, 19/e. pp 268-270.)* Mycoplasmas lack a rigid cell wall and are bound by a triple-layer unit membrane. For this reason, they are completely resistant to the action of penicillins. Unlike the chlamydiae, they can replicate in cell-free media.

299. The answer is C. *(Jawetz, 19/e. pp 268-270.)* Mycoplasmas are extremely small, highly pleomorphic organisms that lack cell walls. They can reproduce on artificial media, where they form small colonies with a "fried egg" appearance. They stain poorly with Gram's stain but well with Giemsa's stain. They are resistant to penicillin but sensitive to tetracycline and sulfonamide.

300. The answer is B. *(Davis, 4/e. pp 693-694.)* Brill-Zinsser disease, or recrudescent typhus, is caused by *R. prowazekii*. The initial episode of louseborne typhus presents with fever, headache, myalgia, and prostration. A macular or maculopapular rash occurs toward the end of the first week. Months to years after the primary infection, a milder disease caused by latent rickettsiae may occur and is known as Brill-Zinsser disease.

301. The answer is E. *(Davis, 4/e. pp 705-706.)* Ornithosis (psittacosis) is caused by *Chlamydia psittaci*. Humans usually contract the disease from infected birds kept as pets or from infected poultry, including poultry in dressing plants. Although ornithosis may be asymptomatic in humans, severe pneumonia can develop. Fortunately, the disease is cured easily with tetracycline.

302. The answer is D. *(Davis, 4/e. pp 696–697.)* Most rickettsial diseases are transmitted to humans by way of arthropod vectors. The only exception is Q fever, which is caused by *Coxiella burnetii*. This organism is transmitted by inhalation of contaminated dust and aerosols or by ingestion of contaminated milk.

303. The answer is D. *(Davis, 4/e. pp 687–697.)* All the listed diseases except Q fever are tick-borne. The rickettsia *Coxiella burnetii* causes Q fever and humans are usually infected by aerosol of a sporelike form shed in milk, urine, feces, or placenta of infected sheep, cattle, or goats. Lyme disease is caused by a spirochete, *Borrelia burgdorferi*, and produces the characteristic lesion erythema chronicum migrans (ECM). The etiologic agent of Rocky Mountain spotted fever is *R. rickettsia*. It usually produces a rash that begins in the extremities and then involves the trunk. Erhlichiosis is caused by *Erhlichia canis*, a rickettsia that is newly recognized as a human pathogen. It was previously considered only a pathogen in dogs. Ehrlichiosis produces the clinical picture in the question. Infection is transmitted by the brown dog tick, produces fever and leukopenia, and usually does not cause a rash. The organism infects monocytes and produces inclusion bodies in the phagosome, where it grows. *Francisella tularensis* is a small, gram-negative, nonmotile coccobacillus. Humans most commonly acquire the organism after contact with tissues or body fluid of an infected mammal or the bite of an infected tick.

304. The answer is A. *(Davis, 4/e. p 693.)* Typhus, spotted fever, and scrub typhus are all caused by rickettsiae (*R. prowazekii*, *R. rickettsii*, and *R. tsutsugamushi*, respectively). Clinically, the diseases have several similarities. Each has an incubation period of 1 to 2 weeks followed by a febrile period, which usually includes a rash. During the febrile period, rickettsiae can be found in the patient's blood and there is disseminated focal vasculitis of small blood vessels. The geographic area associated with these diseases is usually different. Scrub typhus is usually found in Japan, southeast Asia, and the Pacific, while spotted fever is usually found in the western hemisphere. Typhus has a worldwide incidence.

305. The answer is E. *(Mandell, 3/e. pp 1458–1463.)* *Ureaplasma urealyticum* has been associated with nongonococcal urethritis (NGU) as well as infertility. *M. pneumoniae* is the etiologic agent of primary atypical pneumonia. *M. hominis*, although isolated from up to 30 percent of patients with NGU, has yet to be implicated as a cause of that disease. *M. fermentans* has on rare occasions been isolated from the oropharynx and genital tract. *M. mycoides* causes bovine pleuropneumonia.

306. The answer is A. *(Mandell, 3/e. pp 1445–1446.)* Mycoplasmas are the smallest free-living organisms. They can be distinguished from viruses by their growth on cell-free media and from true bacteria by lacking a cell wall. Some species have been shown to be pathogenic for humans.

307. The answer is B. *(Mandell, 3/e. pp 1431–1432.)* The differential diagnosis of lymphogranuloma venereum (LGV) includes syphilis, genital herpes, and chancroid. Several clinical tests can be used to rule out syphilis and genital herpes. These include a positive dark-field examination as well as positive serologic findings for syphilis and the demonstration of herpes simplex virus by cytology or culture. *Haemophilus ducreyi* can usually be isolated from the ulcer in chancroid.

308. The answer is D. *(Mandell, 3/e. pp 1430–1435.)* Trachoma has been the greatest single cause of blindness in the world. *Chlamydia trachomatis* is the most common cause of sexually transmitted disease in the United States and is also responsible for the majority of cases of infant conjunctivitis and infant pneumonia.

309. The answer is B. *(Jawetz, 19/e. pp 306–307.)* LGV is a sexually transmitted disease caused by *Chlamydia trachomatis* of immunotypes L_1, L_2, and L_3. It is more commonly found in tropical climates. In the U.S., the sex ratio is reported to be 3.4 males to 1 female. Tetracycline has been successful in treating this disease in the early stages; however, late stages usually require surgery.

310. The answer is D. *(Jawetz, 19/e. pp 294–298.)* Of the organisms listed in the question, only *Rickettsia rickettsii* penetrates host cells by an active process requiring the expenditure of energy (i.e., ATP). Chlamydiae have a complex growth cycle, which is obligately intracellular. Although the precise mode of penetration is not known, it is likely that a vesicle is formed around the chlamydiae, which then are taken into the cell by a mechanism similar to phagocytosis; chlamydiae do not synthesize ATP. *Mycoplasma* species are free-living bacteria that do not actively penetrate cells.

311. The answer is A. *(Mandell, 3/e. pp 1424–1425.)* The chlamydiae are obligate prokaryotic parasites of eukaryotic cells. For many years they were considered to be viruses but are now considered to be bacteria. The two species, *C. trachomatis* and *C. psittaci*, can be distinguished by two criteria: the susceptibility of *C. trachomatis* to sulfonamides and its ability to form inclusions containing glycogen.

312. The answer is E. *(Jawetz, 19/e. pp 304–305.)* Trachoma is the most common cause worldwide of blindness. It is a chronic keratoconjunctivitis that affects about 400 million people and can be treated with sulfonamides and tetracycline. Relapse of trachoma is common.

313. The answer is B. *(Jawetz, 19/e. pp 300–307.)* The developmental cycle of chlamydiae begins with the "elementary body" attaching to and then penetrating the host cell. The elementary body, now in a vacuole bounded by host-cell membrane, becomes an "initial body." Within about 12 h the initial body has divided to form many small elementary particles encased within an inclusion body in the cytoplasm; these progeny are liberated by host-cell rupture.

314. The answer is C. *(Wilson, 12/e. pp 756–762.)* R. rickettsiae are energy-deficient parasites that cannot use glucose as an energy source without its being phosphorylated. This is thought to be due to a transport defect, rather than to a leaky membrane or atypical cell wall. With the exception of *Rochalimaea quintana*, the agent of trench fever, rickettsiae cannot be cultivated on artificial media. The usual vector for disease is the tick.

315. The answer is D. *(Davis, 4/e. pp 699–703.)* Although both chlamydiae and viruses are obligate intracellular parasites and depend on the host cell for ATP and phosphorylated intermediates, they differ in many respects. Unlike viruses, chlamydiae synthesize proteins, are sensitive to antibiotics, and reproduce by fission. Chlamydiae are readily seen under the light microscope and possess bacterial-like cell walls.

316. The answer is B. *(Davis, 4/e. pp 687–697.)* The etiologic agent of Q fever, *Coxiella burnetii*, is atypical of rickettsiae. It is stable outside the host cell and is resistant to drying. Transmission to humans is by inhalation and not by rodent or arthropod vectors. Rash is not a prominent sign, and Weil-Felix antibodies are not found in the sera of affected persons.

317–320. The answers are: 317-C, 318-E, 319-B, 320-A. *(Jawetz, 19/e. pp 268–270.)* Members of the mycoplasma group that are pathogenic for humans include *Mycoplasma pneumoniae* and *Ureaplasma urealyticum*. *M. pneumoniae* is best known as the causative agent of primary atypical pneumonia (PAP), which may be confused clinically with influenza or legionellosis. It also is associated with arthritis, pericarditis, aseptic meningitis, and the Guillain-Barré syndrome. *M. pneumoniae* can be cultivated on special media and identified by its ability to lyse erythrocytes of sheep or guinea pigs.

Ureaplasma urealyticum (once called *tiny*, or *T, strain*) has been implicated in cases of nongonococcal urethritis. As the name implies, this organism is able to split urea, a fact of diagnostic significance. *U. urealyticum* is part of the normal flora of the genitourinary tract, particularly in women. The only other species of *Mycoplasma* associated with human disease is *M. hominis*. A normal inhabitant of the genital tract of women, this organism has been demonstrated to produce an acute respiratory illness that is associated with sore throat and tonsillar exudate but not with fever.

M. orale and *M. salivarium* are both inhabitants of the normal human oral cavity. These species are commensals and do not play a role in disease.

M. fermentans is an animal isolate.

321–325. The answers are: 321-C, 322-A, 323-B, 324-B, 325-D. *(Jawetz, 19/e. pp 294–298.)* Rickettsiae are small bacteria that are obligate intracellular parasites. Except for *Coxiella* (the agent of Q fever), all rickettsiae are transmitted to humans by arthropods. *Coxiella* is transmitted through the respiratory tract rather than through the skin. In the rare patient, *Coxiella* may cause chronic endocarditis that is not very responsive to either antimicrobial therapy or valve replacement. *Rickettsia rickettsii*, the causative agent of Rocky Mountain spotted fever (RMSF), may be found in wood ticks and is passed via the transovarian route. In the eastern U.S., RMSF is transmitted by the dog tick, *Dermacentor variabilis*.

Rochalimaea quintana is the causative agent of trench fever. The organism can be found in body lice, and humans are the only known reservoir. Disease with this organism was mainly limited to wartime in central Europe, but can be found now in Mexico, Tunisia, Poland, and the former Soviet Union.

Chlamydiae are gram-negative bacteria that are obligate intracellular parasites. They are divided into two species: *C. trachomatis* and *C. psittaci*. Chlamydiae have a unique developmental cycle. The infectious particle is the elementary body. Once inside the cell, the elementary body undergoes reorganization to form a reticulate body. After several replications, the reticulate bodies differentiate into elementary bodies, are released from the host cell, and become available to infect other cells. Three of the 15 serovars of *C. trachomatis* (L_1, L_2, L_3) are known to cause lymphogranuloma venereum (LGV), a sexually transmitted disease.

Mycology

DIRECTIONS: Each question below contains five suggested responses. Select the **one best** response to each question.

326. Although candidiasis of the oral cavity (thrush) usually is controlled by the administration of nystatin, the disseminated or systemic form of candidiasis requires vigorous therapy with

(A) penicillin
(B) amphotericin B
(C) interferon
(D) chloramphenicol
(E) thiabendazole

327. The major cause of favus, a severe form of chronic ringworm of the scalp, is

(A) *Trichophyton schoenleinii*
(B) *Trichophyton rubrum*
(C) *Microsporum canis*
(D) *Malassezia furfur*
(E) *Epidermophyton floccosum*

328. Which of the following pairs of genera are members of the Zygomycetes class and can be seen on microscopic examination to possess rhizoids?

(A) *Absidia* and *Mucor*
(B) *Rhizopus* and *Mucor*
(C) *Rhizopus* and *Absidia*
(D) *Cladosporium* and *Rhizopus*
(E) *Cladosporium* and *Absidia*

329. A slide culture of a dematiaceous mold revealed the image below. The most likely identity of this mold is

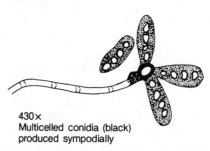

430×
Multicelled conidia (black) produced sympodially

(A) *Drechslera*
(B) *Cladosporium*
(C) *Alternaria*
(D) *Penicillium*
(E) *Acremonium*

330. A rapid method for identification of systematic fungi is the exoantigen technique. Exoantigen methods exist for identification of all the following fungi EXCEPT

(A) *Blastomyces*
(B) *Histoplasma*
(C) *Coccidioides*
(D) *Cryptococcus*
(E) *Paracoccidioides*

331. The object designated by the arrow in the photomicrograph is

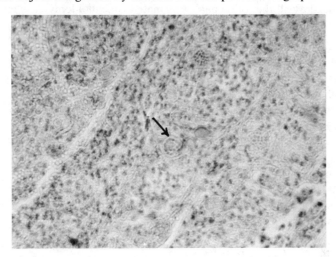

(A) an encapsulated yeast
(B) a thick-walled spore
(C) a spherule
(D) a hyphal strand
(E) a macroconidium

332. A 6-year-old girl presented to the clinic with scaly patches on the scalp. Primary smears and culture of the skin and hair were negative. A few weeks later, she returned and was found to have inflammatory lesions. The hair fluoresced under Wood's light and primary smears of skin and hair contained septate hyphae. On speaking with the parents, it was discovered that there were several pets in the household. Which of the following is the most likely agent?

(A) *Microsporum audouini*
(B) *Microsporum canis*
(C) *Trichophyton tonsurans*
(D) *Trichophyton rubrum*
(E) *Epidermophyton floccosum*

333. An AIDS patient with a persistent cough has shown progressive behavioral changes in the past few weeks after eating an undercooked hamburger. A CSF sample is collected and an encapsulated, yeast-like organism is observed. Based only on these observations, what is the most likely organism?

(A) *Toxoplasma*
(B) *Cryptosporidium*
(C) *Candida*
(D) *Cryptococcus*
(E) *Pneumocystis*

334. In humans, fungal disease can be produced by all the following EXCEPT

(A) invasion of keratin-rich tissues
(B) contamination of wounds with conidia or mycelial fragments
(C) inhalation of conidia
(D) invasion of mucous membranes
(E) ingestion of contaminated food

335. A patient with lymphoma that is being treated with cytotoxic agents has urine, sputum, and blood specimens collected. *Candida albicans* grows from all three. The most effective method to determine whether the *C. albicans* reflects colonization or infection is

(A) detection of mannan antigen in the blood
(B) detection of *Candida* protein antigen by latex agglutination
(C) detection of arabinitol in the blood
(D) assay of IgA antibody to *C. albicans*
(E) isolation of *C. albicans* from three body sites

336. During the third trimester of pregnancy, vaginal infection with which of the following organisms occurs more frequently than normal?

(A) *Candida*
(B) *Acinetobacter*
(C) *Aspergillus*
(D) *Exophiala*
(E) *Epidermophyton*

337. Patients who have disseminated coccidioidomycosis may often demonstrate any of the following EXCEPT

(A) a positive coccidioidin skin test
(B) a negative coccidioidin skin test
(C) immunity to reinfection
(D) a high titer of complement-fixing (CF) antibodies
(E) a negative CF antibody test

338. Which one of the following fungi is the most serious threat in a bone marrow transplant unit?

(A) *Candida albicans*
(B) *Aspergillus*
(C) *Blastomyces*
(D) *Cryptococcus*
(E) *Wangiella*

339. Cryptococci have a polysaccharide capsule with all the following characteristics EXCEPT that it

(A) is an aid to diagnosis
(B) inhibits phagocytosis
(C) cross-reacts with rheumatoid factor
(D) causes a precipitin reaction with hyperimmune rabbit serum
(E) is virulent in the absence of living cryptococci

Mycology

340. A section of tissue from the foot of a person assumed to have eumycotic mycetoma shows a white, lobulated granule composed of fungal hyphae. In the United States, the most common etiologic agent of this condition is a species of

(A) *Acremonium*
(B) *Nocardia*
(C) *Actinomyces*
(D) *Pseudallescheria (Petriellidium)*
(E) *Madurella*

341. The diagnosis of fungal infection may be clinical, serologic, microscopic, or cultural. Although the isolation and identification of a fungus from a suspect lesion establishes a precise diagnosis, it is time-consuming. Microscopy is more rapid but generally less sensitive. Visualization of fungi in a clinical specimen is best accomplished by treatment of the specimen with

(A) silver nitrate
(B) hydrochloric acid
(C) potassium hydroxide
(D) para-aminobenzoic acid
(E) calcofluor white

342. Infection with *Sporothrix schenckii* (formerly *Sporotrichum schenckii*) is an occupational hazard for gardeners. The portal of entry for this organism is the

(A) lymphatic system
(B) respiratory tract
(C) skin
(D) mouth
(E) mucous membranes

343. There are three genera of dermatophytes: *Epidermophyton, Microsporum*, and *Trichophyton*. Infections caused by these organisms (dermatophytoses) are

(A) marked by alveolar irritation
(B) characterized by aflatoxin-induced hallucinations
(C) confined to keratinized tissues
(D) rarely associated with chronic lesions
(E) easily treatable with penicillin

Microbiology

DIRECTIONS: Each group of questions below consists of lettered headings followed by a set of numbered items. For each numbered item select the **one** lettered heading with which it is **most** closely associated. Each lettered heading may be used **once, more than once, or not at all.**

Questions 344–347

Match the fungi listed below with the most appropriate description.

(A) Widespread in environment; conidia may be inhaled; microscopic appearance in specimen reveals dichotomous branching and septate hyphae
(B) Round, black sporangia filled with endospores; sporangia unbranched, rising from a runner called a *stolon*
(C) Single-tipped sporangiophores; no rhizoids or stolons; non-septate hyphae, which show branching
(D) Yeast forms with budding blastoconidia often showing pseudohyphae; positive germ tube test; chlamydospores present
(E) None of the above

344. *Candida albicans*

345. *Aspergillus*

346. *Mucor*

347. *Rhizopus*

Questions 348–352

The naming of fungi is very confusing to the nonmycologist. For this reason, the clinician who may treat fungal infections should have a working knowledge of fungal taxonomy. Most of the fungi known to cause infection in humans have been recognized for many years by their asexual stage (anamorph). The sexual stage (teleomorph) of many of these familiar fungi has now been discovered. Match the numbered anamorph with the appropriate lettered teleomorph.

(A) *Ajellomyces capsulata*
(B) *Ajellomyces dermatitidis*
(C) *Arthroderma van breuseghemii*
(D) *Filobasidiella neoformans*
(E) *Nannizzia incurvata*

348. *Trichophyton mentagrophytes*

349. *Microsporum gypseum*

350. *Cryptococcus neoformans*

351. *Blastomyces dermatitidis*

352. *Histoplasma capsulatum*

Mycology

Questions 353–357

Match the characteristic microscopic findings with the appropriate organism.

(A) *Epidermophyton floccosum*
(B) *Coccidioides immitis*
(C) *Phialophora verrucosa*
(D) *Microsporum canis*
(E) *Blastomyces dermatitidis*

353. Barrel-shaped arthroconidia

354. Sporulation from flask-shaped, pigmented projections

355. Clavate macroconidia

356. Broad-based budding cells

357. Rough-walled macroconidia of 8 to 15 cells

Questions 358–362

For each skin disease below, select the organism most likely to be the causative agent.

(A) *Epidermophyton floccosum*
(B) *Malassezia furfur*
(C) *Microsporum canis*
(D) *Exophiala werneckii*
(E) *Trichosporon beigelii*

358. Tinea corporis

359. Tinea cruris

360. Tinea pedis

361. Tinea capitis

362. Tinea versicolor

Mycology

Answers

326. The answer is B. *(Jawetz, 19/e. pp 323–325.)* Disseminated candidiasis can be life-threatening either as a primary infection in immunosuppressed patients or as a secondary infection of the lungs, kidneys, and other organs in persons who have tuberculosis or cancer. These persons must be treated with amphotericin B or other antifungal drugs. Nystatin, the treatment for candidiasis of the mouth (thrush), does not concentrate in tissues and thus is ineffective in treating disseminated candidiasis.

327. The answer is A. *(Jawetz, 19/e. pp 314–315.)* Trichophyton schoenleinii infection may cause favus, which is characterized by destruction of hair follicles and permanent hair loss. Scutulla (cuplike structures) are formed by crusts around the infected follicles. *Malassezia furfur* causes a fungal skin infection producing brownish-red scaling patches on the neck, trunk, and arms. *Epidermophyton floccosum* and *T. rubrum* are common causes of athlete's foot. *Microsporum canis* infections involve the hair and skin and can be differentiated from *Trichophyton* infections by the ability of *Microsporum* to fluoresce under ultraviolet light.

328. The answer is C. *(Jawetz, 19/e. pp 311–313.)* The *Cladosporium* genus is not in the class Zygomycetes. Members of the *Mucor* genus do not possess rhizoids. The rhizoids of *Absidia* are located on the hyphae between the sporangiophores; the rhizoids of *Rhizopus* are located directly beneath the sporangiophores.

329. The answer is A. *(Balows, Clinical Microbiology, 5/e. pp 644–658.)* Drechslera is a dematiaceous fungus that had been previously named *Helminthosporium*. Colonies are fluffy and gray to brownish black in color. The hyphae are septate and the conidia are multi-septate and elongate. The conidiophores may be twisted.

330. The answer is D. *(Balows, Clinical Microbiology, 5/e. pp 635–636.)* Dimorphic fungi are those that are in the yeast phase at 37°C and the mycelium phase at 27 to 30°C. These fungi are isolated from patients in the yeast phase and must be converted to the mycelium phase before definitive identification can take place. Most of the systematic fungi except *Coccidioides*—

that is, *Histoplasma, Blastomyces, Paracoccidioides,* and *Sporothrix*—can be converted by a combination of selective agar and temperature. Exoantigen techniques in which specific antigens are identified by immunodiffusion have greatly improved both the speed and ease of identification of these fungi. Although rapid methods exist for the detection of *Cryptococcus* in body fluids, they are not strictly exoantigen techniques but are based on detection of the specific cryptococcal polysaccharide capsule with antibody-sensitized latex particles.

331. The answer is B. *(Jawetz, 19/e. pp 325–327.)* Thick-walled spores are characteristic of many fungal infections, including blastomycosis, coccidioidomycosis, and histoplasmosis. Observation of these structures in sputum or in tissue should alert the microbiologist to a diagnosis of systemic fungal infection. The presence of encapsulated yeast in clinical specimens may suggest the presence of *Cryptococcus*.

332. The answer is B. *(Jawetz, 19/e. pp 314–316.)* Hairs infected with *Microsporum canis* and *M. audouini* both fluoresce with a yellow-green color under Wood's light, while *Trichophyton rubrum, T. tonsurans,* and *Epidermophyton floccosum* do not. But *M. audouini* is an anthropophilic agent of tinea capitis, whereas *M. canis* is zoophilic. *M. canis* is primarily seen in children and is associated with infected cats or dogs.

333. The answer is D. *(Jawetz, 19/e. pp 325–327.)* Patients with paralysis of their cellular immune system, such as in AIDS, are susceptible to a wide variety of diseases, including infection with *Cryptococcus*. A brain abscess caused by *Cryptococcus neoformans* is not unusual in AIDS patients. Initial laboratory suspicion is usually aroused by the presence of encapsulated yeast in the CSF. There also could be other microorganisms as well as noninfectious artifacts that superficially resemble yeast. While *C. neoformans* can be readily cultured, a rapid diagnosis can be made by detecting cryptococcal capsular polysaccharide in CSF or blood. Care must be taken to strictly control the test because rheumatoid factor may cross-react. Once the yeast is isolated, then specific stains as well as panels of assimilatory carbohydrates are available to definitively identify this organism as *C. neoformans*. The patient may also be infected with *Pneumocystis carinii*, but not in the central nervous system. *P. carinii* has recently been reclassified as a fungus.

334. The answer is E. *(Davis, 4/e. pp 746–748.)* Cutaneous mycoses (dermatophytoses) and superficial mycoses cause disease in skin, hair, and nails by invasion of keratinized tissue. Systematic fungal disease is caused by inhalation of conidia. Infections due to direct implantation of conidia in the skin, subcutaneous mycoses, are caused by soil saprophytes. Fungi of normal flora can directly invade a susceptible host through mucous membranes and cause local or disseminated disease (vulvovaginal candidiasis in pregnant or diabetic women, for example). If fungal food poisoning occurs, it is a rare event. However, aspergilli produce toxins that may contaminate foods such as peanut butter and flour.

335. The answer is E. *(Balows, Clinical Microbiology, 5/e. pp 617–629.)* The determination of whether a patient is colonized or infected with *C. albicans* is not straightforward. Usually, isolation of the yeast from three sites is indicative of infection. Determination of antibody titer has little clinical utility. Detection of mannan or arabinitol has correlated with infection in some studies but not in others. Recent evidence suggests that detection of the protein antigen by latex agglutination may not signify disseminated candidiasis.

336. The answer is A. *(Jawetz, 19/e. pp 323–325.)* Candida, and in particular *C. albicans*, may be found as part of the normal flora of the mouth, vagina, and gastrointestinal tract. As a pathogen, it is an opportunistic fungus. When invasive, candidiasis can be an acute or chronic infection, either localized or disseminated. High levels of glucose in the blood of women who are in the third trimester of pregnancy or who have diabetes encourage vulvovaginal candidiasis.

337. The answer is E. *(Jawetz, 19/e. pp 318–320.)* In patients with coccidioidomycosis, a positive skin test to coccidioidin appears 2 to 21 days after the appearance of disease symptoms and may persist for 20 years without reexposure to the fungus. A decrease in intensity of the skin response often occurs in clinically healthy people who move away from endemic areas. A negative skin test frequently is associated with disseminated disease. Complement-fixing (CF) immunoglobulin G (IgG) antibodies, which may not appear at all in mild disease, rise to a high titer in disseminated disease, a poor prognostic sign. For this reason, a persistent or rising CF titer combined with clinical symptoms indicates present or imminent dissemination. Rarely is the CF titer negative. Most persons infected with *Coccidioides immitis* are immune to reinfection.

338. The answer is B. *(Balows, Clinical Microbiology, 5/e. pp 661–662.)* While all fungi are potentially serious in a bone marrow transplant unit, the most frequent cause of fungal infection and death is *Aspergillus*. Aspergilli probably originate from the environment. There are instances of multiple infections in new units that have not been monitored prior to opening or in units adjacent to construction projects.

Mycology Answers

339. The answer is E. *(Jawetz, 19/e. pp 325–327.)* The characteristic capsules of cryptococci allow the yeast cells to be seen easily in India ink suspensions. All strains of cryptococci produce capsules. Only *Cryptococcus neoformans* is pathogenic in humans; it is found in those who have a weak immune response to this organism. Hyperimmunized rabbits, however, produce capsule-specific antisera that differentiate among three strains of *C. neoformans*. The precipitin reaction with rabbit serum is the only serologic test for antibodies of diagnostic value. Latex tests for cryptococcal antigen may be falsely positive owing to a cross reaction with rheumatoid factor. Rheumatoid factor may be removed by heating or by enzymatic digestion. The capsule is not directly toxic or virulent if injected into experimental animals.

340. The answer is D. *(Jawetz, 19/e. pp 317–318.)* Eumycotic mycetoma is a slowly progressing disease of the subcutaneous tissues that is caused by a variety of fungi. The term *Madura foot* has been used to describe the foot lesion. Although several fungi have been isolated in the United States from persons who have mycetoma, *Pseudallescheria boydii* appears to be one of the most common. Other foot infections that may resemble Madura foot are actinomycotic (bacterial) in nature. These are caused by *Nocardia brasiliensis* and *Actinomadura*.

341. The answer is E. *(Howard, pp 536–541.)* Evidence supporting the presence of fungal infection includes the clinical appearance of lesions and positive serologic reactions. However, detection of fungi in lesions, either by microscopic inspection or by culture, is the best evidence. Treating a specimen with 10% sodium hydroxide or potassium hydroxide hydrolyzes protein, fat, and many polysaccharides, leaving the alkali-resistant cell walls of most fungi intact and visible. With the availability of calcofluor white, however, KOH is not the optimum method for detecting and differentiating the structures of fungal cell walls. Calcofluor white is a nonimmunologic fluorescent stain that selectively stains fungi. Additionally, fluorescent antibody stains are available for the rapid diagnosis of fungal infection.

342. The answer is C. *(Howard, p 578.)* Cutaneous sporotrichosis, caused by *Sporothrix schenckii*, begins at the site of inoculation, usually on an extremity or the face. The organism often is found on thorns of rose bushes. Ulceration is common and new lesions appear along paths of lymphatic channels. Extracutaneous sporotrichosis is seen primarily in bones and joints. There is no evidence to suggest that any portal of entry besides skin is important.

343. The answer is C. *(Jawetz, 19/e. pp 314–316.)* The dermatophytes are a group of fungi that infect only superficial keratinized tissue (skin, hair, nails). They form hyphae and arthroconidia on the skin; in culture, they develop colonies and conidia. Tinea pedis, or athlete's foot, is the most common dermatophytosis. Several topical antifungal agents, such as undecylenic acid, salicylic acid, and ammoniated mercury, may be useful in treatment. For serious infection, systemic use of griseofulvin is effective.

344–347. The answers are: 344-D, 345-A, 346-C, 347-B. *(Jawetz, 19/e. pp 323–327.)* Fungi that cause opportunistic infections are diverse, and most of them are represented in this group of questions. Infection occurs primarily in the compromised host with underlying diseases such as lymphoma, leukemia, and diabetes. Unfortunately, most of the opportunistic fungi that cause infection are commonly seen in the laboratory as contaminants.

Candidiasis is the most frequent opportunistic infection. While *C. albicans* is most commonly isolated, other species such as *C. tropicalis* and *Torulopsis glabrata* are also seen. The yeasts may be identified biochemically, but *C. albicans* is distinctive in that it produces germ tubes and chlamydospores.

Zygomycosis, a term referring to infection by members of the class Zygomycetes, is caused by *Rhizopus*, *Mucor*, and *Absidia* primarily. Other Zygomycetes such as *Basidiobolus* and *Cunninghamella* are rarely encountered. The lack of septated hyphae on a direct smear may be the initial hint of zygomycosis. However, not uncommonly, the occasional hypha of *Mucor* will have a septa. The genera cannot be differentiated on a direct patient specimen. The organism must be isolated and slide cultures performed to observe the characteristic morphology of these filamentous fungi.

Rhizopus species have sporangia that arise from a stolon, while *Mucor* species do not. *Mucor* species have collarettes; *Rhizopus* species do not.

Aspergillosis, caused by a number of species of *Aspergillus*, is characterized in direct smear by septated hyphae, dichotomously branched. *A. flavus* and *A. fumigatus* are often seen as saprophytes in the laboratory but also account for the major species isolated from patients with aspergillosis. Differentiation of species, as with the Zygomycetes, is dependent upon isolation of the fungus and precise morphologic examination.

348–352. The answers are: 348-C, 349-E, 350-D, 351-B, 352-A. *(Howard, pp 528–530.)* The classification of fungi is complicated because one pathogenic fungus (holomorph) may have two names—that of the anamorph (asexual form) and that of the teleomorph (sexual form). For example, the teleomorph of *Histoplasma capsulatum* is *Ajellomyces capsulata*. It is also interesting to note that the teleomorph (*Ajellomyces*) of two distinct genera, *Blastomyces* and *Histoplasma*, is the same. Similarly the dermatophyte *Microsporum gyp-*

seum is the anamorph of two distinctly different sexual forms—*Nannizzia gypsea* and *N. incurvata*—and the teleomorph of *Trichophyton mentagrophytes* is *Arthroderma van breuseghemii*. The commonly known pathogenic fungus *Cryptococcus neoformans* has as its teleomorph *Filobasidiella neoformans*, a name that to date has little clinical meaning.

For those fungi in which no sexual stage has been found, the term *fungi imperfecti* serves as a convenient repository of asexual forms. In clinical practice, to avoid confusion, the name of the asexual stage is routinely reported.

353-357. The answers are: 353-B, 354-C, 355-A, 356-E, 357-D. *(Jawetz, 19/e. pp 311-331.)* Microscopic examination of fungal isolates is essential to the identification of the organism. Macroscopically, the colonies of *Epidermophyton* have a yellowish appearance. This fungus invades skin and nails but never hair. On microscopic examination, clavate or paddle-shaped macroconidia are evident with rounded ends and smooth walls. Microconidia are absent.

Coccidioides immitis is a dimorphic fungi endemic in some regions of the southwestern U.S. and Latin America. In tissue, the organism exists as a spherule filled with endospores. When grown on solid media, the organism produces barrel-shaped arthroconidia, which stain with lactophenol cotton blue.

Phialophora verrucosa is one of the causes of chromoblastomycosis, a chronic localized infection of the skin and subcutaneous tissue. Microscopically, short or somewhat elongated, flask-shaped pigmented phialides are seen. The collarettes are vase-shaped and darkly pigmented.

Microsporum canis is a dermatophyte that infects skin and hair but rarely nails. When hair is infected with this organism, it will fluoresce. Microscopic examination of this organism demonstrates rough-walled macroconidia of 8 to 15 cells.

Blastomyces dermatitidis causes a chronic granulomatous disease. The yeast cells are globose or ovoid in shape. The single blastoconidium is attached by a broad base to the parent cell.

358-362. The answers are: 358-C, 359-A, 360-A, 361-C, 362-B. *(Davis, 4/e. pp 760-765.)* Dermatomycoses are cutaneous mycoses caused by three genera of fungi: *Microsporum, Trichophyton,* and *Epidermophyton*. These infections are called *tinea* or *ringworm*, a misnomer that has persisted from the days when they were thought to be caused by worms or lice.

Tinea capitis (ringworm of the scalp) is due to an infection with *M. canis* or *T. tonsurans*. It usually occurs during childhood and heals spontaneously at puberty. Circular areas on the scalp with broken or no hair are characteristic of this disorder.

Tinea corporis (ringworm of the body) is caused by *M. canis* and *T. mentagrophytes*. This disorder affects smooth skin and produces circular pruritic areas of redness and scaling.

Both tinea cruris (ringworm of the groin, "jock itch") and tinea pedia (ringworm of the feet, "athlete's foot") are caused by *T. rubrum, T. mentagrophytes,* or *E. floccosum*. These common conditions are pruritic and can cause scaling.

Tinea versicolor (pityriasis versicolor) is not a dermatomycotic condition but, rather, a superficial mycosis now thought to be caused by *Malassezia furfur*. The disorder is characterized by chronic but asymptomatic scaling on the trunk, arms, or other parts of the body.

Parasitology

DIRECTIONS: Each question below contains five suggested responses. Select the **one best** response to each question.

363. Members of the genus *Babesia* are sporozoan parasites of the red blood cell that have recently been found to cause human infections. Infection usually results in a febrile illness associated with hemolysis. Humans become infected from

(A) ticks
(B) mosquitoes
(C) fleas
(D) mites
(E) none of the above

364. A patient who had extended-wear contact lenses complained to an ophthalmologist about increasing irritation of the eye. The physician sent the patient's contact lens cleaning solution to the laboratory. A wet mount revealed many ameboid organisms. Without further diagnostic or laboratory investigation, the most likely identification of the organism in the lens solution is

(A) *Acanthamoeba*
(B) *Hartmannella*
(C) *Pneumocystis*
(D) *Naegleria*
(E) *Paramecium*

365. The single most common cause of pneumonia in AIDS patients is

(A) *Streptococcus pneumoniae*
(B) *Cryptosporidium*
(C) *Mycobacterium tuberculosis*
(D) *Pneumocystis carinii*
(E) *Toxoplasma*

366. Assuming that a variety of microorganisms might cause pneumonia in AIDS patients, which of the following combinations of stains would optimally differentiate the major causes of the pneumonia if a piece of lung tissue was available?

(A) Gram's stain and methylene blue stain
(B) Direct fluorescent antibody stain and acid-fast stain
(C) Acid-fast stain and Albert's stain
(D) Methylene blue and calcifluor white stain
(E) Lactophenol cotton blue and Giemsa stain

367. The parasite shown in the blood smear commonly remains in the lymphatic system in the early stages of infestation. This organism, which is the causative agent of elephantiasis, is

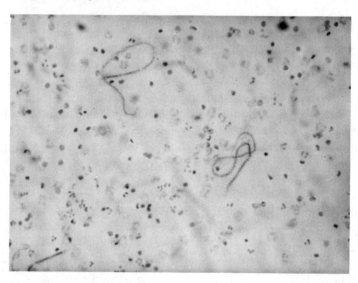

(A) *Strongyloides stercoralis*
(B) *Ancylostoma duodenale*
(C) *Ascaris lumbricoides*
(D) *Wuchereria bancrofti*
(E) *Trichuris trichiura*

368. In order to exert control over the primary cause of toxoplasmosis of pregnancy, which one of the following steps of the life cycle of *Toxoplasma* would be most practical to interrupt?

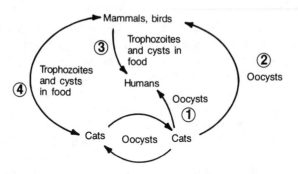

(A) Step 1
(B) Step 2
(C) Step 3
(D) Step 4
(E) Steps 3 and 4

---

369. All the following statements concerning Lyme disease are true EXCEPT

(A) at least half of patients with Lyme disease do not develop erythema around the tick bite
(B) there are different species of *Ixodes* ticks that carry *Borrelia burgdorferi* depending on the location of the infestation in the U.S.
(C) *Borrelia*-infected ticks are restricted to the genus *Ixodes*
(D) dogs may be bitten by ticks carrying *Borrelia* and become infected
(E) white-tailed deer are an important reservoir for *Ixodes dammini* nymphal ticks

370. A previously healthy 14-year-old boy was admitted to the hospital with severe frontal and bitemporal headache, lethargy, and fever. During the 3 weeks prior to admission, he had been swimming and diving in a freshwater lake. A lumbar puncture was done and examination revealed an elevated white blood cell count, primarily polymorphonuclear leukocytes, and motile amebae. The organism is most likely to be

(A) *Entamoeba histolytica*
(B) *Entamoeba polecki*
(C) *Dientamoeba fragilis*
(D) *Acanthamoeba*
(E) *Naegleria fowleri*

371. The diagnostic characteristics of *Plasmodium falciparum* are best described by which one of the following statements?

(A) A period of 72 h is required for the development of the mature schizont, which resembles a rosette with only 8 to 10 oval merozoites
(B) An important diagnostic feature is the irregular appearance of the edges of the infected red blood cell
(C) The signet-ring–shaped trophozoite is irregular in shape with ameboid extensions of the cytoplasm
(D) Except in infections with very high parasitemia, only ring forms of early trophozoites and the gametocytes are seen in the peripheral blood
(E) Schüffner stippling is routinely seen in red blood cells that harbor parasites

372. A Marine Corps sergeant who recently returned from Operation Desert Storm in Saudi Arabia reported to "sick bay" with cutaneous lesions on his lips and cheeks. A Giemsa stain of the lesions revealed darkly staining kinetoplasts and light-staining nuclei within macrophages. The most likely cause of these lesions is

(A) *Trypanosoma*
(B) *Toxoplasma*
(C) *Sarcocystis*
(D) *Histoplasma*
(E) *Leishmania*

373. Tapeworms capable of producing human infection in their larval stages include all the following EXCEPT

(A) hydatid cyst of *Echinococcus*
(B) cercocystis (cysticercoid) of *Hymenolepis nana*
(C) sparganum of *Diphyllobothrium*
(D) cysticercus of *Taenia solium*
(E) cysticercus of *Taenia saginata*

374. Which of the following techniques is employed most successfully for recovering pinworm eggs?

(A) Sugar fecal flotation
(B) Zinc-sulfate fecal flotation
(C) Tap-water fecal sedimentation
(D) Direct fecal centrifugal flotation
(E) Anal swabbing with cellophane tape

375. Human infection with the beef tapeworm, *Taenia saginata*, usually is less serious than infection with the pork tapeworm, *T. solium*, because

(A) acute intestinal stoppage is less common in beef tapeworm infection
(B) larval invasion does not occur in beef tapeworm infection
(C) toxic by-products are not given off by the adult beef tapeworm
(D) the adult beef tapeworms are smaller
(E) beef tapeworm eggs cause less irritation of the mucosa of the digestive tract

376. The finding of eosinophilia in the peripheral blood is

(A) unaffected by host response
(B) not noted in nonparasitic forms of infection
(C) an invaluable diagnostic indicator of parasitic infection
(D) more marked in recent than in chronic parasitic infection
(E) indicative of a systemic parasitic disease, such as parasitemia

377. A man coughed up a long (4 to 6 cm) white worm and his chief complaint is abdominal tenderness. He reports that he goes to sushi bars at least once a week. The following parasites have been observed in people who eat raw fish: *Anisakis*, *Pseudoterranova*, *Eustrongylus*, and *Angiostrongylus*. Which of the following would best differentiate the specific parasitic agent?

(A) Identification of specific species of fish involved
(B) Study of distinctive morphology of the parasite
(C) Specific antibody tests
(D) Antigen detection in tissues
(E) Characteristic signs and symptoms

378. Recommendations for the control of human hookworm in endemic areas include the construction of sanitary facilities and the

(A) thorough washing of fresh fruit and vegetables
(B) thorough cooking of all meats
(C) reduction of the wild dog population
(D) use of insecticides to control flies
(E) wearing of footwear

379. In the typical life cycle of a trematode (e.g., *Schistosoma*), which of the following developmental forms enters the intermediate snail host?

(A) Cercaria
(B) Metacercaria
(C) Schizont
(D) Redia
(E) Miracidium

380. All the following statements describe human lice EXCEPT

(A) they are wingless
(B) they cause pruritic skin lesions
(C) they transmit epidemic typhus, relapsing fever, and trench fever
(D) *Pediculus humanus* and *Phthirus pubis* are two species
(E) they secrete a potent neurotoxin

381. A survey of 100 healthy adults reveals that 80 percent have IgG antibodies to *Toxoplasma*. Which one of the following statements would help to explain this finding?

(A) The potential for *Toxoplasma* infection is widespread and the disease is mild and self-limiting
(B) Toxoplasmosis is caused by eating meat; therefore, all meat-eaters have had toxoplasmosis
(C) A variety of parasitic infections induce the formation of *Toxoplasma* antibody
(D) The test for *Toxoplasma* antibodies is highly nonspecific
(E) The IgM test is more reliable than the IgG test for determination of past infections; retesting for IgM would show that most people do not have *Toxoplasma* antibody

382. One million persons in the United States have roundworm infection. Which of the following parasites is a roundworm that hatches in the upper small intestine and releases rhabditiform larvae that penetrate the intestinal wall?

(A) *Hymenolepis nana*
(B) *Diphyllobothrium latum*
(C) *Schistosoma mansoni*
(D) *Fasciola hepatica*
(E) *Ascaris lumbricoides*

383. Organisms of the type depicted below are seen in diarrheic feces. This finding is most compatible with a diagnosis of

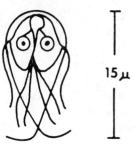

(A) bilharziasis
(B) ascariasis
(C) enterobiasis
(D) giardiasis
(E) shigellosis

384. Scabies is caused by a small mite that burrows into the skin. The disease is described by all the following statements EXCEPT

(A) it is caused by a species of *Sarcoptes*
(B) it is often complicated by secondary bacterial infection
(C) it is synonymous with Kawasaki syndrome
(D) it is best diagnosed by morphologic identification of the mite
(E) allergic (asthma-like) reactions to mites are often reported

385. In the United States, certain enteric protozoan and helminthic infections were previously considered to be exotic illnesses related to foreign travel or to contaminated food or water. However, sexual transmission of these diseases has produced a "hyperendemic" infection rate among male homosexuals. The most common infection seen in this group is

(A) giardiasis
(B) ascariasis
(C) amoebiasis
(D) enterobiasis
(E) trichiuriasis

386. Analysis of a patient's stool reveals small structures resembling rice grains; microscopic examination shows these to be proglottids. The most likely organism in this patient's stool is

(A) *Enterobius vermicularis*
(B) *Ascaris lumbricoides*
(C) *Necator americanus*
(D) *Taenia saginata*
(E) *Trichuris trichiura*

387. An AIDS patient complains of headaches and disorientation. A clinical diagnosis of *Toxoplasma* encephalitis is made. Which one of the following antibody results would be most likely in this patient?

(A) IgM nonreactive, IgG nonreactive
(B) IgM nonreactive, IgG reactive (low titer)
(C) IgM reactive (low titer), IgG reactive (high titer)
(D) IgM reactive (high titer), IgG reactive (high titer)
(E) IgM reactive (high titer), IgG nonreactive

388. Most tapeworms are intestinal parasites of humans. Humans can be both the intermediate and the definitive host of

(A) *Taenia solium*
(B) *Taenia saginata*
(C) *Diphyllobothrium latum*
(D) *Dirofilaria immitis*
(E) *Echinococcus granulosus*

389. *Trypanosoma cruzi* initially penetrates through the mucous membranes on the skin and then multiplies in a lesion known as a *chagoma*. In the chronic stage of the disease, the main lesions are often observed in the

(A) spleen and pancreas
(B) heart and digestive tract
(C) liver and spleen
(D) digestive tract and respiratory tract
(E) heart and liver

390. The species of the protozoan *Trichomonas* that are found in humans are *T. hominis* and *vaginalis*. Which one of the following statements is true?

(A) *T. hominis* is the most pathogenic for humans
(B) *T. hominis* infects the distal small intestine
(C) *T. vaginalis* usually causes erosion of the uterine mucosa
(D) *T. vaginalis* is transmitted sexually
(E) There are no morphologic differences between *T. hominis* and *T. vaginalis*

391. A woman recently returned from Africa complains of having paroxysmal attacks of chills, fever, and sweating; these attacks last a day or two at a time and recur every 36 to 48 h. Examination of a stained blood specimen reveals ringlike and crescentlike forms within red blood cells. The infecting organism most likely is

(A) *Plasmodium falciparum*
(B) *Plasmodium vivax*
(C) *Trypanosoma gambiense*
(D) *Wuchereria bancrofti*
(E) *Schistosoma mansoni*

Questions 392–393

A young man, recently returned to the United States from Viet Nam, has severe liver disease. Symptoms include jaundice, anemia, and weakness.

392. The etiologic agent, shown in the photomicrographs on the next page, is

(A) *Plasmodium falciparum*
(B) *Clonorchis sinensis*
(C) *Diphyllobothrium latum*
(D) *Taenia solium*
(E) *Taenia saginata*

393. An intermediate form of the organism shown in the photomicrographs on the next page lives in

(A) mosquitoes
(B) pigs
(C) snails
(D) cows
(E) ticks

Photomicrographs accompany Questions 392–393

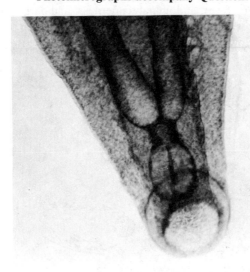

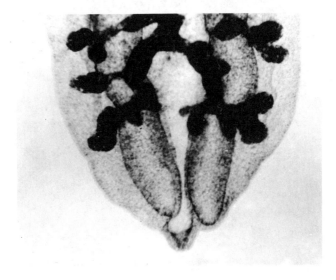

394. A woman who recently traveled through Central Africa now complains of severe chills and fever, abdominal tenderness, and darkening urine. Her febrile periods last for 28 h and recur regularly. Which of the blood smears drawn below would be most likely to be associated with the symptoms described?

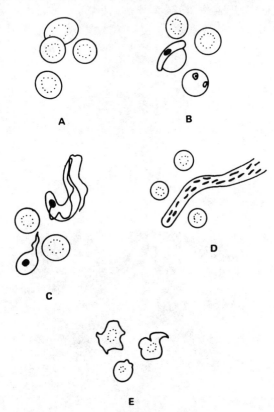

(A) A
(B) B
(C) C
(D) D
(E) E

395. One of the most clinically significant infections in patients with AIDS is *Pneumocystis carinii* pneumonia (PCP). PCP is a treatable disease; therefore, rapid diagnosis is essential. The method of choice for detection of *P. carinii* in respiratory specimens is

(A) methenamine silver stain
(B) toluidine blue stain
(C) direct fluorescent antibody (DFA) microscopy
(D) indirect fluorescent antibody (IFA) microscopy
(E) culture in rat lung cells

396. There are five varieties of cockroaches: the German cockroach, the brown-handed cockroach, the oriental cockroach, the American cockroach, and the smoky brown cockroach. A characteristic of cockroaches is their

(A) transmission of *Salmonella*
(B) toxic sting
(C) function as a vector for *Borrelia burgdorferi*
(D) function as a secondary host for rickettsiae
(E) easy eradication

397. Amebae that are parasitic in humans are found in the oral cavity and the intestinal tract. All the following statements about intestinal amebae are true EXCEPT

(A) they are often nonpathogenic
(B) they can cause peritonitis and liver abscesses
(C) they are usually transmitted as cysts
(D) they occur most abundantly in the cecum
(E) infection with *Entamoeba histolytica* is limited to the intestinal tract

398. Schistosomiasis is a disease characterized by granulomatous reactions to the ova or to products of the parasite at the place of oviposition. Clinical manifestations include all the following EXCEPT

(A) calcifications in the bladder wall
(B) pulmonary arterial hypertension
(C) splenomegaly
(D) esophageal varices
(E) arthropathies

399. The photomicrograph below shows fine fibrils (labeled "F") in an ameba. These structures are

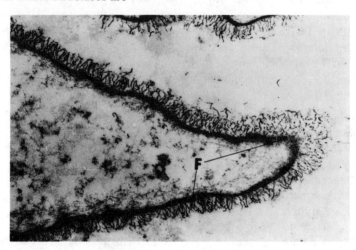

(A) termed *amebic microtubules*
(B) analogous to actin in the cells of higher forms of life
(C) primarily of glycoprotein composition
(D) not involved in cell motility
(E) inactive at 37°C (98.6°F)

Parasitology

DIRECTIONS: Each group of questions below consists of lettered headings followed by a set of numbered items. For each numbered item select the **one** lettered heading with which it is **most** closely associated. Each lettered heading may be used **once, more than once, or not at all.**

Questions 400–404

Parasitic trematodes of humans are defined, according to their location in the host, as blood, liver, intestinal, or lung flukes. Match each fluke with its appropriate primary intermediate host, secondary intermediate host, and final habitat.

(A) Snails, watercress, bile ducts
(B) Snails, aquatic plants, liver
(C) Snails, fish, bile ducts
(D) Snails, crabs, vesical plexus
(E) Snails, no secondary intermediate host, mesenteric veins

400. *Schistosoma mansoni*

401. *Schistosoma japonicum*

402. *Fasciola hepatica*

403. *Clonorchis sinensis*

404. *Opisthorchis felineus*

Questions 405–408

A primary procedure for diagnosis of fecal parasites is a stained smear of feces. For some parasitic infections, however, other specimens may be more productive. Match each parasite with the best method of identification.

(A) Sigmoidoscopy and aspiration of mucosal lesions
(B) Baermann technique
(C) Dilution followed by egg count
(D) Enzyme immunoassay (EIA)
(E) Examination of a cellophane tape swab

405. *Giardia lamblia*

406. *Entamoeba histolytica*

407. *Strongyloides* larvae

408. *Ascaris*

Questions 409–413

Match the case histories below with the most likely diagnosis.

(A) Trichinosis
(B) Schistosomiasis
(C) Toxoplasmosis
(D) Visceral larva migrans
(E) Giardiasis

409. A butcher, who is fond of eating raw hamburger, develops chorioretinitis; a Sabin-Feldman dye test is positive

410. A fur trapper complains of sore muscles, has swollen eyes, and reports eating bear meat on a regular basis

411. A newspaper correspondent has diarrhea for 2 weeks following a trip to St. Petersburg (Leningrad)

412. A retired Air Force colonel has had abdominal pain for 2 years; he makes yearly freshwater fishing trips to Puerto Rico and often wades with bare feet into streams

413. A teenager who works in a dog kennel after school has had a skin rash, eosinophilia, and an enlarged liver and spleen for 2 years

Parasitology

Answers

363. The answer is A. *(Ash, 3/e. p 115.)* *Babesia microti* is a species of sporozoa in New England that has caused both sporadic and epidemic infections in humans. These parasites appear as small rings within the red blood cell—an appearance similar to that of *Plasmodium falciparum*. Unlike *Plasmodium* infections, *Babesia* infections have no associated pigment in the red cell. The vector is the tick *Ixodes dammini*. This tick also harbors *Borrelia burgdorferi*, the agent of Lyme disease, and *Erlichia canis*, the agent of erlichiosis seen in both dogs and humans.

364. The answer is A. *(Balows, Clinical Microbiology, 5/e. pp 744–748.)* Contaminated contact lens solutions have been recently implicated in serious infections of the eye caused by *Acanthamoeba*. The use of homemade saline is to be avoided; most commercial solutions, however, have preservatives in them. The organisms are difficult to isolate, but the differential criteria for the free-living ameba are both nutritional and morphologic.

365. The answer is D. *(Balows, Clinical Microbiology, 5/e. pp 742–744.)* One of the multiple criteria for classification of AIDS is the development of *Pneumocystis carinii* pneumonia (PCP). *Pneumocystis* is a fungus formerly thought to be a parasite that was classified with the sporozoa. In addition to AIDS, PCP may also be seen in patients with congenital or other acquired cellular immune dysfunction. Most AIDS patients are given prophylactic aerosolized pentamidine for PCP. PCP is easily diagnosed in respiratory secretions using a direct fluorescent antibody test.

366. The answer is B. *(Balows, Clinical Microbiology, 5/e. pp 323–330.)* *Pneumocystis carinii*, *Mycobacterium tuberculosis*, and *Mycobacterium avium-intracellulare* are the respiratory pathogens most commonly seen in patients with AIDS. The diagnosis of these infections is best made from tissue. Minimally, the tissue sections should be stained with direct fluorescent antibody stains for *P. carinii* and an acid-fast stain for *Mycobacterium*. Methylene blue is a nonspecific stain. While the Gram's stain (bacteria), lactophenol cotton blue stain (fungi), and Giemsa stain (parasites, multinucleated giant cells) are all useful, none would be the first choice for AIDS patients. A DNA probe based on the polymerase chain reaction (PCR) is available in research laboratories for *P. carinii* and *Mycobacterium* in sputum.

Microbiology

367. The answer is D. *(Ash, 3/e. pp 168–171.)* Filariasis, also known as elephantiasis and wuchereriasis, is caused by the extensive growth and proliferation of *Wuchereria bancrofti*. The parasite is transmitted through the bite of mosquitoes, and humans are the only definitive hosts. Microfilariae are common in blood and can be observed in blood smears from infected persons.

368. The answer is A. *(Balows, Clinical Microbiology, 5/e. pp 740–742.)* *Toxoplasma gondii* may be acquired by inhalation of oocysts in cat feces. It is difficult to control the habits of cats unless they are housed and not let out. Pregnant females should avoid changing cat litter boxes. While ingestion of oocysts in raw meat may also lead to toxoplasmosis, inhalation of oocysts is the primary cause, particularly among pregnant women in the United States.

369. The answer is C. *(Balows, Clinical Microbiology, 5/e. pp 560–566.)* In the U.S. *Borrelia burgdorferi*, the causative agent of Lyme disease, has two principal vectors: *Ixodes dammini* in the eastern and midwestern U.S. and *I. pacificus* in the western U.S. The ticks are tiny and can easily be missed. Unfortunately, only 50 percent or less of patients develop significant inflammation around the tick bite. Lyme disease, usually with joint involvement, is also seen in veterinary patients such as dogs, cats, and horses. White-tailed deer and small rodents are an important reservoir for these ticks. *B. burgdorferi* has been isolated from *Dermacentor* and *Amblyomma* ticks and from mosquitoes as well as from several *Ixodes* species.

370. The answer is E. *(Ash, 3/e. pp 65–67.)* Primary amebic meningoencephalitis, associated with swimming in freshwater lakes, is a rare, usually fatal disease. The causative agent, *Naegleria fowleri*, gains access to the central nervous system by being forced under pressure through the nasal mucosa covering the cribriform plate. Diagnosis is made by observation of the motile amebae in a wet mount of the cerebrospinal fluid. The organism can be cultured, and the indirect fluorescent antibody test can confirm the identification. Despite the fact that most cases are fatal, one patient has responded to amphotericin B alone and another patient to amphotericin B, miconazole, and rifampin. Although *Acanthamoeba* has been reported as a cause of uveitis and corneal ulceration, a central nervous system syndrome has been induced in laboratory animals. In humans, the clinical illness is poorly defined.

371. The answer is D. *(Balows, Clinical Microbiology, 5/e. pp 727–736.)* *P. falciparum* infection is distinguished by the appearance of ring forms of early trophozoites and gametocytes, both of which can be found in the peripheral blood. The size of the RBC is usually normal. Double dots in the rings are common.

Parasitology Answers

372. The answer is E. *(Balows, Clinical Microbiology, 5/e. pp 738–739.)* There have been a number of cases of both visceral and cutaneous leishmaniasis in returning Desert Storm veterans. The CDC has recommended that all returning veterans not donate blood, although the risk of transfusion-related leishmaniasis appears low. The kinetoplast aids in differentiating amastigotes from other similar organisms that may be found in macrophages.

373. The answer is E. *(Balows, Clinical Microbiology, 5/e. pp 779–780.)* The adult worm of *Echinococcus* lives in the small intestine of dogs and certain wild animals. Humans acquire their parasites from feces of infected dogs. The larval stage of *Taenia solium* is called *Cystericercus cellulosae*. Humans usually acquire this disease by the ingestion of food or water contaminated by infected human feces. Humans can acquire sparganosis by the ingestion of infected *Cyclops* in drinking water or by consuming infected frogs, snakes, or rodents. *Hymenolepis nana*, the dwarf tapeworm, is distributed worldwide and may be contracted by ingestion of infected insects. Eggs of *T. saginata* (beef tapeworm) are not directly infective for humans, but caution should be used in handling gravid proglottids.

374. The answer is E. *(Ash, 3/e. pp 130–133.)* The pinworm *Enterobius vermicularis* is a parasite of the cecum and intestine. At night, female worms migrate to deposit eggs in perianal and perineal regions. Eggs can be recovered easily by perianal swabbing with cellophane tape; the eggs adherent to the tape can be identified microscopically. Swabbing for 3 consecutive days reveals the presence of eggs in 90 percent of cases. The search for pinworm eggs in fecal specimens is often unrewarding.

375. The answer is B. *(Ash, 3/e. pp 221–237.)* Both beef tapeworm (*Taenia saginata*) and pork tapeworm (*T. solium*) can, in the adult form, cause disturbances of intestinal function. Intestinal disorder is due not only to direct irritation but also to the action of metabolic toxic wastes. In addition, *T. saginata*, because of its large size, may produce acute intestinal blockage. Unlike *T. saginata*, *T. solium* produces cysticercosis, which results in serious lesions in humans (in *T. saginata*, the cysticercus—encysted larvae—stage develops only in cattle).

376. The answer is D. *(Wilson, 12/e. pp 774–776.)* Too much emphasis often is placed on eosinophilia as a definite sign of parasitic disease. Eosinophilia is found in a variety of diseases, including asthma, gonorrhea, varicella-zoster, periarteritis nodosa, carcinoma of the rectum, and many others. Eosinophilia is not a consistent sign of parasitic disease; the degree of eosinophilia varies with both host response and duration of infection (it is more marked in recent than in chronic parasitic infection).

377. The answer is B. *(Balows, Clinical Microbiology, 5/e. pp 775–777.)* The consumption of raw fish products in oriental restaurants, especially the growing popularity of sushi and sashimi, has led to a variety of infections, most of which are characterized by symptoms consistent with intestinal blockage or meningitis. The parasites are tissue nematodes and parasites of marine mammals. Fish, squid, and other edible marine life are often secondary hosts. The most reliable way to differentiate the specific helminth is by examination of the whole worm or by histologic examination of the parasites in tissue sections.

378. The answer is E. *(Ash, 3/e. pp 146–147.)* Necator americanus and Ancylostoma duodenale are responsible for most human hookworm infections. Filariform larvae of these organisms gain access to humans by penetrating skin, usually of the interdigital spaces between the toes of people who are barefoot. Warm climate, presence of fecal matter, and damp, loosely packed soil are ideal conditions for the growth and spread of hookworms.

379. The answer is E. *(Ash, 3/e. pp 208–211.)* In the typical life cycle of trematodes, eggs are discharged from the intestinal or genitourinary tract of a definitive host. The eggs hatch in freshwater, releasing the larval miracidia, which enter the snails that serve as intermediate hosts. By metamorphosis, miracidia become rediae, which in turn develop into cercariae. The cercariae are released from the intermediate host and reenter the water. To cause human infection, encysted metacercariae must be ingested; on the other hand, cercariae can penetrate skin. The schizont is an asexual form of malarial protozoa and is not a developmental form of trematodes.

380. The answer is E. *(Balows, Clinical Microbiology, 5/e. pp 799–801.)* Pediculus humanus (head or body louse) and Phthirus pubis (crab louse) are wingless parasites exclusively affecting humans. Lice are important not only for the itching and discomfort they cause but also for the diseases they transmit. These disorders include epidemic typhus, relapsing fever, and trench fever. There is no evidence that lice secrete toxins.

381. The answer is A. *(Balows, Clinical Microbiology, 5/e. pp 740–742.)* Serologic tests, such as the Sabin-Feldman dye test and indirect immunofluorescence, have shown that a high percentage of the world's population has been infected with *Toxoplasma gondii*. In adults, clinical toxoplasmosis usually presents as a benign syndrome resembling infectious mononucleosis. However, fetal infections are often severe and associated with hydrocephalus, chorioretinitis, convulsions, and death. Acute toxoplasmosis is best diagnosed by an IgM capture assay. In most patients, specific IgM antibody disappears within 3 to 6 months.

Parasitology Answers

382. The answer is E. *(Ash, 3/e. pp 134–137, 216–219.)* Helminths are subdivided into three phyla: the Annelida, or segmented worms; the Nemathelminthes, or roundworms; and the Platyhelminthes, or flatworms. *Ascaris lumbricoides* is a roundworm that hatches in the upper small intestine of infected humans; rhabditiform larvae are released and penetrate the intestinal wall. The other four parasites listed in the question are flatworms: *Diphyllobothrium latum* and *Hymenolepis nana* are cestodes (tapeworms), and *Fasciola hepatica* and *Schistosoma mansoni* are trematodes (flukes).

383. The answer is D. *(Ash, 3/e. pp 74–77.)* The organism sketched in the question is too small to be a worm and too large to be a bacterium. It is the trophozoite form of *Giardia lamblia*. Giardiasis can cause acute diarrhea, abdominal pain, and weight loss. It is spread through contaminated food and water.

384. The answer is C. *(Balows, Clinical Microbiology, 5/e. pp 807–809.)* *Sarcoptes scabiei* is a small mite that burrows into human skin. Itching is significant, and a vesicular eruption, which often becomes secondarily infected with bacteria, develops. Diagnosis is made by microscopic detection of the mites. Gamma benzene hexachloride (Kwell), a topical insecticide, is an effective treatment for scabies. A number of etiologic agents have been proposed for Kawasaki disease (KD), among them mites, but there is no evidence that mites either cause or are vectors of KD.

385. The answer is A. *(Ash, 3/e. pp 74–77.)* The infection rate with *Giardia lamblia* in male homosexuals has been reported to be from 21 to 40 percent. These high prevalence rates are probably related to three factors: the endemic rate, the sexual behavior that facilitates transmission (the usual barriers to spread have been interrupted), and the frequency of exposure to an infected person.

386. The answer is D. *(Ash, 3/e. pp 130–133, 138–139, 221–225.)* *Enterobius* (pinworm), *Ascaris* (roundworm), *Necator* (hookworm), and *Trichuris* (whipworm) are roundworms, or nematodes. *Taenia saginata* (tapeworm), a segmented flatworm, affects the small intestine of humans. Tapeworm segments, called *proglottids*, appear in the stool of infected persons.

387. The answer is B. *(Balows, Clinical Microbiology, 5/e. pp 742–743.)* One of the leading causes of death among AIDS patients is central nervous system toxoplasmosis. It is thought that *Toxoplasma* infection is a result of reactivation of old or preexisting toxoplasmosis. Occasionally, the infection may be acquired by needle-sharing. Because the disease is a reactivation of old or preexisting toxoplasmosis, routine quantitative tests for IgM antibody are usually negative and IgG titers are low (<1:256, IFA). More sophisticated methods, such as IgM capture or IgG avidity, may reveal an acute response.

388. The answer is A. *(Ash, 3/e. pp 226–229, 231–237.)* The definitive host is defined as the host that allows full development of the parasite into the adult form. The intermediate host permits the penetration into tissue by the larval stage and then survival in that tissue for varying periods of time. Humans are definitive hosts and acceptable intermediate hosts for *Taenia solium* and *Hymenolepis nana.*

389. The answer is B. *(Ash, 3/e. pp 118–121.)* American trypanosomiasis (Chagas' disease) is produced by *T. cruzi*, which is transmitted to humans by the bite of an infected reduviid bug. After multiplication, the tissues most likely to be affected in the chronic stage of the disease are the cardiac muscle fibers and the digestive tract. A diffuse interstitial fibrosis of the myocardium results and may lead to heart failure and death. The inflammatory lesions in the digestive tract that are seen in the esophagus and colon produce considerable dilatation. Chagas' disease has not been an important disease in the U.S.; most cases have been imported, although there are a few reports of endogenous disease in the southern U.S.

390. The answer is D. *(Balows, Clinical Microbiology, 5/e. pp 760–761.)* Two species of *Trichomonas* are known to inhabit humans. *T. hominis* is found in the intestine, and *T. vaginalis* is found in the genitourinary tract. *T. hominis* is considered nonpathogenic, whereas *T. vaginalis* is responsible for vaginal (but not uterine) infections in women and prostatic and urethral infections in men. Transmission of *T. vaginalis* occurs during coitus. Trichomoniasis in both women and men can be cured by the administration of metronidazole (Flagyl). These trichomonads are differentiated on the basis of motility, shape, number of nuclei, and number and location of flagella.

391. The answer is A. *(Ash, 3/e. pp 106–113.)* The febrile paroxysms of *Plasmodium malariae* malaria occur at 72-h intervals; those of *P. falciparum* and *P. vivax* malaria occur every 48 h. The paroxysms usually last 8 to 12 h with *P. vivax* malaria but can last 16 to 36 h with *P. falciparum* disease. In *P. vivax, P. ovale,* and *P. malariae* infections, all stages of development of the organisms can be seen in the peripheral blood; in malignant tertian (*P. falciparum*) infections, only early ring stages and gametocytes are usually found.

392. The answer is B. *(Ash, 3/e. pp 196–213.)* The Chinese liver fluke, *Clonorchis sinensis,* is a parasite of humans that is found in Japan, China, Korea, Taiwan, and Indochina. Humans usually are infected by eating uncooked fish.

Parasitology Answers

The worms invade bile ducts and produce destruction of liver parenchyma. Anemia, jaundice, weakness, weight loss, and tachycardia may follow. Treatment is likely to be ineffectual in heavy infections, but chloroquine can destroy some of the worms.

393. The answer is C. *(Ash, 3/e. pp 196–213.)* The life cycle of *Clonorchis sinensis* is similar to that of other trematodes. A mollusk is characteristically the first intermediate host of trematodes. For *C. sinensis*, snails perform this role.

394. The answer is B. *(Ash, 3/e. pp 106–113.)* The case history presented in the question is characteristic of infection with *Plasmodium falciparum*, the causative agent of malignant tertian malaria. The long duration of the febrile stage rules out other forms of malaria. The presence of ringlike young trophozoites and crescentlike mature gametocytes—as represented in the illustration below—as well as the absence of schizonts is diagnostic of *P. falciparum* malaria.

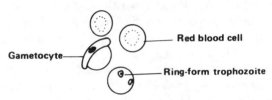

395. The answer is C. *(Balows, Laboratory Diagnosis, pp 959–968.)* Both methenamine silver and toluidine blue stain pneumocysts nonspecifically. These preparations are difficult to read because background material may nonspecifically stain black or blue. *P. carinii* cannot be routinely cultured from human specimens. Both IFA and DFA tests are FDA-approved and available for detection of *P. carinii*. The advantage of DFA is that it is quicker (45 to 60 min versus 3 h) and there is less nonspecific fluorescence observed in the preparation.

396. The answer is A. *(Balows, Clinical Microbiology, 5/e. pp 797–798.)* Cockroaches are nuisances and multiply rapidly in homes, hospitals, and factories. While sprays may be effective, roaches often hide in places not reached by sprays. The biggest public health problem with roaches is that they may carry *Salmonella* and contaminate food or surfaces that they contact.

397. The answer is E. *(Balows, Clinical Microbiology, 5/e. pp 752–754.)* Of the intestinal amebae, *Entamoeba hartmanni*, *E. coli*, *E. polecki*, and *E. nana* are considered nonpathogenic. *E. histolytica* is distinctively characterized by its pathogenic potential for humans, although infection with this protozoan is commonly asymptomatic (causing "healthy carriers"). Symptomatic amebiasis and dysentery occur when the trophozoites invade the intestinal wall and produce ulceration and diarrhea. Peritonitis can occur, with the liver the most common site of extraintestinal disease. The life cycle of the ameba is simple. There is encystment of the "troph," followed by excystation in the ileocecal region. The trophs multiply and become established in the cecum, where encystation takes place and results in abundant amebae, cysts, and trophozoites. Infection is spread by the cysts, which can remain for weeks or months in appropriately moist surroundings.

398. The answer is E. *(Ash, 3/e. pp 208–215.)* Although the chronic stage of proliferation within tissues is distinctive in the different forms of schistosomiasis, a granulomatous reaction to the eggs and chemical products of the schistosome occurs in all forms of the disease. *Schistosoma haematobium* commonly involves the distal bowel and the bladder, as well as the prostate gland and seminal vesicles. Bladder calcification and cancer may ensue. *S. mansoni* affects the large bowel and the liver; presinusoidal portal hypertension, splenomegaly, and esophageal varices may be complications. Pulmonary hypertension, often fatal, may be seen with *S. mansoni* and *S. japonicum* disease. Eggs may be found in an unstained specimen of rectal mucosa or in stool. Urine microscopy and liver biopsy, where warranted, often prove positive. Schistosomiasis is best prevented by the elimination of the parasite in snails before human infection occurs.

399. The answer is B. *(Ash, 3/e. pp 46–55.)* The movement of amebic trophozoites is usually unidirectional and controlled in part by chemotactic factors in the immediate environment. Amebae are motile by virtue of pseudopods, which are cytoplasmic extensions that alternately project and contract. The fibrils shown in the amebic pseudopod pictured in the question are involved in cellular motility and are similar to muscle actin. Amebic motility is retarded as environmental temperature falls below 37°C (98.6°F).

400–404. The answers are: 400-E, 401-E, 402-A, 403-C, 404-C. *(Ash, 3/e. pp 208–215.)* Three species of blood flukes infect humans—*Schistosoma mansoni, S. japonicum,* and *S. haematobium*. Humans are the principal definitive hosts, and a snail is the primary intermediate host. For each species, there is a specific snail host but no secondary host. In *S. japonicum* and *S. mansoni* infection, the mature worms are found in the mesenteric veins.

Fasciola hepatica, Clonorchis sinensis, and *Opisthorchis felineus* are liver flukes. Snails are the primary intermediate host for all. The cercariae of

F. hepatica produced in the snail encyst as metacercariae on watercress. Humans are an incidental host; the larvae penetrate the intestinal wall into the peritoneum, pass through the liver capsule and tissues, and establish themselves in the branches of the biliary duct. The secondary intermediate host for both *C. sinensis* and *O. felineus* is the fish. Humans are the incidental host; the adult worms mature inside the bile ducts.

405–408. The answers are: 405-D, 406-A, 407-B, 408-C. *(Balows, Clinical Microbiology, 5/e. pp 751–769.)* It is not uncommon that repeated stool specimens do not reveal the suspected parasite. Also, microscopic analysis of stool may not reveal parasite load when such data are necessary. For these reasons, other techniques are available to identify parasites as well as to quantitate them.

During sigmoidoscopy, a curette or suction device may be used to scrape or aspirate material from the mucosal surface. Cotton swabs should not be used. A direct mount of this material should immediately be examined for *E. histolytica* trophozoites and then a permanent stain made for subsequent examination.

The Baermann technique may be helpful in recovering *Strongyloides* larvae. Essentially, fecal material is placed on damp gauze on the top of a glass funnel that is three-quarters filled with water. The larvae migrate through the damp gauze and into the water. The water may then be centrifuged to concentrate the *Strongyloides*.

Worm burdens may be estimated by a number of microscopic methods. While not often done, such procedures may provide data on the extent of infection or the efficacy of treatment of hookworms, *Ascaris,* or *Trichuris.* Thirty thousand *Trichuris* eggs per gram, 2000 to 5000 hookworm eggs per gram, and 1 *Ascaris* egg are clinically significant and suggest a heavy worm burden.

The diagnosis of giardiasis is usually made by detecting trophozoites and cysts of *Giardia lamblia* in consecutive fecal specimens. Alternatively, a gelatin capsule on a string (Enterotest) can be swallowed, passed to the duodenum, and then retrieved after 4 h. The string is then examined for *Giardia*. A recent innovation is the introduction of an EIA for *G. lamblia*. The EIA is more sensitive than microscopy, can be performed on a single stool specimen, and does not depend on the presence of entire trophozoites and cysts.

A cellophane tape swab is used to trap pinworms crawling out of the anus during the night. The tape is then examined microscopically for *Enterobius*.

409–413. The answers are: 409-C, 410-A, 411-E, 412-B, 413-D. *(Balows, Clinical Microbiology, 5/e. pp 717–719, 722–723, 774.)* All the diseases listed in the question have significant epidemiologic and clinical features. Toxoplasmosis, for example, is generally a mild, self-limiting disease; however, severe fetal disease is possible if pregnant women ingest *Toxoplasma* oocysts. Consumption of uncooked meat may result in either an acute toxoplasmosis or a chronic toxoplasmosis that is associated with serious eye disease. Most adults have antibody titers to *Toxoplasma* and thus would have a positive Sabin-Feldman dye test.

Trichinosis most often is caused by ingestion of contaminated pork products. However, eating undercooked bear, walrus, raccoon, or possum meat also may cause this disease. Symptoms of trichinosis include muscle soreness and swollen eyes.

Although giardiasis has been classically associated with travel in Russia, especially St. Petersburg (Leningrad), many cases of giardiasis caused by contaminated water have been reported in the United States as well. Diagnosis is made by detecting cysts in the stool. In some cases, diagnosis may be very difficult because of the relatively small numbers of cysts present. Alternatively, an enzyme immunoassay may be used to detect *Giardia* antigen in fecal samples.

Schistosomiasis is a worldwide public-health problem. Control of this disease entails the elimination of the intermediate host snail and removal of stream-side vegetation. Abdominal pain is a symptom of schistosomiasis.

Visceral larva migrans is an occupational disease of people who are in close contact with dogs and cats. The disease is caused by the nematodes *Toxocara canis* (dogs) and *T. cati* (cats) and has been recognized in young children who have close contact with pets or who eat dirt. Symptoms include skin rash, eosinophilia, and hepatosplenomegaly.

Immunology

DIRECTIONS: Each question below contains five suggested responses. Select the **one best** response to each question.

414. It is determined an infant suffers from Bruton's agammaglobulinemia. Which of the following pathogens will present the most serious threat to this child?

(A) Measles virus
(B) *Mycobacterium tuberculosis*
(C) *Streptococcus pneumoniae*
(D) *Chlamydia trachomatis*
(E) Varicella-zoster virus

415. All the following statements concerning immunoglobulin structure are true EXCEPT

(A) the amino acid sequence variation of the heavy chains is similar to that observed in light chains
(B) in man, there are approximately twice as many Ig molecules with kappa and lambda chains
(C) in the three-dimensional structure of Ig, there is little, if any, flexibility in the hinge region between the Fc and two Fab portions
(D) IgM is a pentameric structure
(E) myeloma proteins have been widely used for Ig structural studies

416. Which of the following statements is true concerning natural killer (NK) cells?

(A) They belong to T-cell lineage
(B) They belong to B-cell lineage
(C) They kill bacterially infected cells
(D) They require prior antigen exposure for activation
(E) They display cytotoxic effect on tumor cells

417. Which class of immunoglobulin is associated with anaphylactic delayed hypersensitivity reaction?

(A) IgA
(B) IgD
(C) IgE
(D) IgG
(E) IgM

418. Bence Jones proteins, which are often found in the urine of persons who have multiple myeloma, are best described as

(A) mu chains
(B) gamma chains
(C) kappa and lambda chains
(D) albumin derivatives
(E) fibrin split products

419. Specific immunologic unresponsiveness is called *tolerance*. All the following statements regarding tolerance are true EXCEPT

(A) it is related to the immunologic maturity of the host
(B) it occurs only with polysaccharide antigens
(C) it is related to the dose of antigen
(D) it is best maintained by the presence of antigen
(E) it is prolonged by administration of immunosuppressive drugs

420. IgA is the main immunoglobulin in secretions of the respiratory, intestinal, and genital tracts. All the following statements regarding IgA are true EXCEPT

(A) secretory IgA has a molecular weight of 400,000
(B) IgA can be found in saliva and tears
(C) some bacteria can destroy IgA via specific protease production
(D) IgA can fix complement
(E) some IgA exists in serum as a monomer

421. Which of the following is an immune abnormality often seen in patients with AIDS?

(A) Increased CD8+ T cells
(B) Decreased CD4+ T cells
(C) CD4+/CD8+ ratios greater than 2/1
(D) Hypogammaglobulinemia
(E) Absence of secretory IgA

422. It appears that HIV binds selectively to CD4 glycoproteins. Thus, HIV shows a selective infection and destruction of helper T cells. All the following cells exhibit CD4 glycoprotein on their cell surface EXCEPT

(A) macrophages
(B) neurons
(C) glial cells
(D) vaginal epithelial cells
(E) squamous epithelial cells

423. A young girl has had repeated infections with *Candida albicans* and respiratory viruses since the time she was 3 months old. As part of the clinical evaluation of her immune status, her responses to routine immunization procedures should be tested. In this evaluation, the use of which of the following vaccines is contraindicated?

(A) Diphtheria toxoid
(B) *Bordetella pertussis* vaccine
(C) Tetanus toxoid
(D) BCG
(E) Inactivated polio

424. All the following describe properties of interleukin 1 (IL-1) EXCEPT

(A) it is a macrophage-derived product
(B) it may activate B cells
(C) it may stimulate cytotoxic T cells
(D) three biologically active forms are described
(E) its activity is histocompatibility-restricted

425. A patient, recently divorced, complained of being tired and "run down" for the past few months. On initial screening, her "Monospot" (infectious mononucleosis slide agglutination test) was negative. All the following tests might be helpful in making the diagnosis of chronic infectious mononucleosis EXCEPT

(A) EBV-VCA-IgG (EBV viral capsid antigen)
(B) EBV-VCA-IgM
(C) EBV-VCA-IgA
(D) EBNA (EBV nuclear antigen)
(E) EA (early antigen)

426. All the following statements regarding interleukin 1 (IL-1) are true EXCEPT

(A) it can be produced by activated macrophages
(B) it can be produced by natural killer cells
(C) it exerts its effects on T and B cells as a costimulator
(D) it consists of a single protein
(E) it acts synergistically with IL-6

427. The hinge region of an IgG heavy chain is located

(A) between V_H and C_{H1}
(B) between C_{H1} and C_{H2}
(C) between C_{H2} and C_{H3}
(D) within the C_{H1} intrachain disulfide loop
(E) within the Fc fragment

428. The Arthus reaction is a classic inflammatory response that is best described by which of the following statements?

(A) The Arthus reaction requires a low concentration of antigen and antibody
(B) The Arthus reaction appears later after injection than does passive cutaneous anaphylaxis
(C) The Arthus reaction is mediated by immunoglobulin M
(D) The characteristic Arthus lesion develops slowly
(E) The extent of the Arthus lesion is independent of the quantity of reacting antigen and antibody

429. Survival of allografts is increased by choosing donors with few major histocompatibility (MHC) mismatches with recipients and by use of immunosuppression in recipients. All the following might be useful measures of immunosuppression EXCEPT

(A) administration of corticosteroids to recipient
(B) lymphoid irradiation of recipient
(C) administration of antilymphocyte globulin to recipient
(D) destruction of donor B cells
(E) destruction of activated T cells

430. A latent, measleslike viral infection and, presumably, a defect in cellular immunity is associated with which of the following diseases?

(A) Progressive multifocal leukoencephalopathy (PML)
(B) Multiple sclerosis
(C) Creutzfeldt-Jakob disease
(D) Subacute sclerosing panencephalitis
(E) Epstein-Barr virus infection

431. Which of the following subclasses of antibody is most abundant in serum?

(A) IgG-1
(B) IgG-2
(C) IgG-3
(D) IgG-4
(E) IgA-1

432. In humans, two closely linked genetic loci, each made up of two alleles, compose the histocompatibility locus A (HL-A). Paired first and second locus antigens are called *haplotypes*. The HL-A haplotypes (separated by a semicolon) of a child's parents are given below.

      Father 3,25; 7,12
      Mother 1,3; 8,9

Assuming that no cross-over events have occurred, the child's histotype could be which of the following?

(A) 1,3; 7,8
(B) 7,12; 1,3
(C) 3,3; 7,9
(D) 1,25; 7,12
(E) 3,25; 7,12

433. All the following statements about macrophages are true EXCEPT

(A) they are activated by CD4+ T cells
(B) they secrete interleukin 1
(C) lipopolysaccharide (LPS) does not activate macrophages
(D) lysosomal enzymes are produced by macrophages
(E) they can exhibit MHC-II proteins

434. The amounts of protein precipitated in a series of tubes containing a constant amount of antibody and varying amounts of antigen are presented below. In which tube is antigen-antibody equivalence obtained?

Tube	Antigen (mg)	Protein precipitated (mg)
(A)	0.02	1.1
(B)	0.08	2.1
(C)	0.32	3.1
(D)	1.0	3.7
(E)	2.0	2.9

Immunology

435. The graph below shows the sequential alteration in the type and amount of antibody produced after an immunization. (Inoculation of antigen occurs at two different times, as indicated by the arrows.) Curve A and curve B each represents a distinct type of antibody. The class of immunoglobulin represented by curve B has which of the following characteristics?

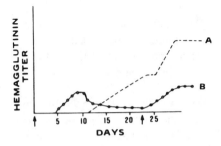

(A) An estimated molecular weight of 150,000
(B) A composition of four peptide chains connected by disulfide links
(C) An appearance in neonates at approximately the third month of life
(D) The human ABO isoagglutinin
(E) A symmetric dipeptide

436. Which of the statements about the precipitin curve shown below is true?

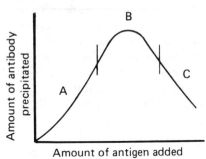

(A) In a multispecific system, a solution in zone B would have only an excess of antigen in the supernatant
(B) In a monospecific system, a solution in zone B would contain only reacted antibody and antigen
(C) A solution in zone A would be expected to have unreacted precipitable antigen in the supernatant
(D) A solution in zone C would be expected to have both antigen and antibody in excess
(E) A solution in zone C would be expected to have an excess of antibody in the supernatant

437. Thrombocytopenic patients who have aplastic anemia benefit from platelet transfusions during periods of severe platelet depression. Platelets may be rejected because of an ABO or HLA incompatibility. Which of the following platelet-transfusion donors would be best for a patient whose blood type is A− and who has HLA haplotypes 1,3; 7,12?

(A) O−   1,3; 7,12
(B) O−   3,5; 7,8
(C) AB−   2, −; 5,13
(D) A+   1,3; 7,12
(E) A+   3,5; 7,8

438. Which of the following T-cell markers can be found on human T cells?

(A) CD8
(B) Ia
(C) QA1
(D) Ly1
(E) Immunoglobulin

439. Which one of the following immunoglobulin classes can activate complement by the alternative pathway?

(A) IgA
(B) IgD
(C) IgE
(D) IgG
(E) IgM

440. Relative to the primary immunologic response, secondary and later booster responses to a given hapten-protein complex can be associated with all the following EXCEPT

(A) higher titers of antibody
(B) increased antibody affinity for the hapten
(C) increased antibody avidity for the original hapten-protein complex
(D) a shift in subclass, or idiotype, of antibody produced
(E) antibodies that are less efficient in preventing specific disease

441. An Ouchterlony gel diffusion plate shows the reaction of a polyspecific serum against several antigen preparations. The center well in Figure 1 below contains polyspecific antiserum, first bleed; the center well in Figure 2 contains polyspecific antiserum, second bleed; NS is normal saline. In this situation, cross reaction can be recognized between antigen X and antigen

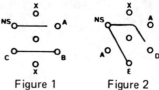

Figure 1   Figure 2

(A) A
(B) B
(C) C
(D) D
(E) E

Immunology

Questions 442–444

The Scatchard plot shown below represents the interaction of a hapten molecule with an immunoglobulin in an equilibrium dialysis apparatus. This interaction is defined by the equation

$$K = \left(\frac{r}{n-r}\right)c \text{ or } \frac{r}{c} = Kn - Kr$$

(K is the intrinsic affinity constant, c is the free concentration of hapten, r is the number of hapten molecules bound per antibody molecule at c, and n is the antibody valence.)

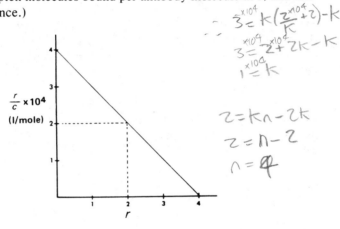

442. The affinity constant for this system is

(A) 1×10^{-4} moles/liter
(B) 1×10^{4} moles/liter
(C) 1×10^{4} liters/mole
(D) -4×10^{-4} moles/liter
(E) 16×10^{-4} moles/liter

443. The antibody valence (n) is defined as the maximum number of ligand molecules able to be bound per antibody molecule. In the example presented, n equals

(A) 1
(B) 2
(C) 3
(D) 4
(E) 10

444. The antibody species most likely to have been used in the experiment described is

(A) IgA
(B) IgD
(C) IgE
(D) IgG
(E) IgM

445. Transfer factor has the capacity to transfer cell-mediated responses to a nonreactive recipient. Which of the following statements about transfer factor is true?

(A) It is resistant to heating (56°C, 30 min)
(B) It has a molecular weight of 200,000
(C) It is sensitive to RNase and DNase
(D) It is inactivated by trypsin
(E) It is derived from human leukocytes

446. All the following hypotheses may be sufficient to explain nonprecipitation in an antigen-antibody system EXCEPT

(A) the antigen has a monovalent determinant
(B) the antigen has multiple, closely repeated determinants
(C) the antibody has been cleaved to divalent Fab' ligands
(D) the antibody has been cleaved to divalent $F(ab')_2$ ligands
(E) the antibody has low affinity for the antigen

Immunology

DIRECTIONS: The group of questions below consists of four lettered headings followed by a set of numbered items. For each numbered item select

A	if the item is associated with	(A) **only**
B	if the item is associated with	(B) **only**
C	if the item is associated with	**both** (A) and (B)
D	if the item is associated with	**neither** (A) nor (B)

Each lettered heading may be used **once, more than once, or not at all.**

Questions 447–451

(A) B-cell function
(B) T-cell function
(C) Both
(D) Neither

447. Skin test for delayed-type hypersensitivity

448. Serum immunoglobulin level

449. Lymphocyte proliferation induced by concanavalin A

450. White blood cell count

451. Biopsy examined for plasma cells by histologic technique

DIRECTIONS: Each group of questions below consists of lettered headings followed by a set of numbered items. For each numbered item select the **one** lettered heading with which it is **most** closely associated. Each lettered heading may be used **once, more than once, or not at all.**

Questions 452–456

For each of the descriptions below, match the immunologic technique.

(A) Affinity chromatography
(B) Radial immunodiffusion
(C) Rocket immunoelectrophoresis
(D) Western or immunoblot
(E) Enzyme-linked immunosorbent assay (ELISA)

452. Antigen undergoes electrophoresis through polyacrylamide gel and is transferred to nitrocellulose, where it is reacted with antisera

453. Antigen is placed in a well of agar containing uniformly dispersed antiserum

454. Antigen is absorbed to a solid phase and allowed to react; interaction is detected by covalently linked enzyme-Ab complexes, which react with substrates yielding colored products

455. Antiserum is passed through a column of agarose beads to which antigen or hapten is coupled

456. Antigen undergoes electrophoresis from a well into agar that contains an excess of antisera

Questions 457–461

Match the laboratory or clinical observations below with the appropriate clinical syndrome or microbial etiology.

(A) Fifth disease
(B) Susceptibility to chickenpox
(C) Possible subacute sclerosing panencephalitis (SSPE)
(D) Possible hepatitis B infection
(E) Acute Lyme disease

457. IgM antibody (1:200) to *Borrelia burgdorferi*

458. Elevated IgG and IgM antibody titers to parvovirus

459. Negative varicella antibody titer

460. Increased antibody titer to delta agent

461. Elevated CSF antibody titer to measles virus

Immunology

Questions 462–465

Infection with Epstein-Barr virus (EBV) results in the development of virus-specific antibodies. In the case of infectious mononucleosis, for each antibody listed below, choose the description with which it is most likely to be associated.

(A) Appears 2 weeks to several months after onset and is present more often in atypical cases of infectious mononucleosis
(B) Appears 3 to 4 weeks after onset; titers correlate with severity of clinical illness
(C) Arises early in the course of the illness; detectable titers persist a lifetime
(D) Appears late in the course of the disease and persists a lifetime
(E) Arises early in the course of the illness, and then titers fall rapidly

462. EBNA-Ab

463. EA-EBV (anti-D) Ab

464. EBV-VCA (IgG) Ab

465. EBV-VCA (IgM) Ab

Questions 466–470

For each disease listed below, choose the level of immune function (humoral and cellular) with which it is most likely to be associated.

	Humoral	Cellular
(A)	Normal	Normal
(B)	Normal	Deficient
(C)	Deficient	Normal
(D)	Deficient	Deficient
(E)	Elevated	Elevated

466. Ataxia-telangiectasia

467. Infantile X-linked agammaglobulinemia (Bruton's disease)

468. Swiss-type hypogammaglobulinemia

469. Thymic hypoplasia (DiGeorge's syndrome)

470. Wiskott-Aldrich syndrome

Questions 471–474

For each of the terms below, select the best description.

(A) Transplant from one region of a person to another
(B) Transplant from one person to a genetically identical person
(C) Transplant from one person to a genetically different person
(D) Transplant from one species to another species
(E) None of the above

471. Xenograft

472. Allograft

473. Autograft

474. Isograft

Questions 475–477

Antigenic determinants on immunoglobulins are used to classify antibodies. For each antibody classification below, select the determinant with which it is most likely to be associated.

(A) Determinant exposed after papain cleavage to an $F(ab')_2$ fragment
(B) Determinant from one clone of cells and probably located close to the antigen-binding site of the immunoglobulin
(C) Determinant inherited in a Mendelian fashion and recognized by cross-immunization of individuals in a species
(D) Heavy-chain determinant recognized by heterologous antisera
(E) Species-specific carbohydrate determinant on the heavy chain

475. Isotype

476. Allotype

477. Idiotype

Immunology

Questions 478–482

For each diagnosis given, choose the serum electrophoretic profile with which it is most likely to be associated.

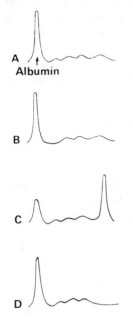

478. α_1-Antitrypsin deficiency

479. Multiple myeloma

480. Swiss-type agammaglobulinemia

481. Polyclonal hypergammaglobulinemia

482. Normal

Questions 483–485

Complement-fixation (CF) testing is an important serologic tool. For each reaction mixture below, select the expected result.

(A) Complement is bound, red blood cells are lysed
(B) Complement is bound, red blood cells are not lysed
(C) Complement is not bound, red blood cells are lysed
(D) Complement is not bound, red blood cells are not lysed
(E) Complement is not bound, red blood cells are agglutinated

483. Anti-*Mycoplasma* antibody + complement + hemolysin-sensitized red blood cells (RBC) + anti-RBC antibody

484. Anti-*Mycoplasma* antibody + *Mycoplasma* antigen + complement + hemolysin-sensitized red blood cells

485. Anti-*Mycoplasma* antibody + *Mycoplasma* antigen + complement + hemolysin-sensitized red blood cells + anti-RBC antibody

Questions 486–490

Match the tests below with the diseases for which they are diagnostic.

(A) Infectious mononucleosis
(B) *Mycoplasma pneumoniae* pneumonia
(C) Rubella
(D) Syphilis
(E) *Staphylococcus aureus* endocarditis

486. Teichoic acid antibodies

487. Hemagglutination inhibition

488. Complement fixation

489. Viral capsid antigen (VCA) IgG or IgM antibodies

490. Rapid plasma reagin (RPR)

Questions 491–495

Most, but not all, cases of hepatitis are caused by hepatitis A virus (HAV), hepatitis B virus (HBV), or non-A, non-B hepatitis virus. While the laboratory diagnosis of HAV is usually accomplished by the detection of IgG and IgM antibodies to HAV, the diagnosis of HBV is more complex. Match the immunologic markers with the appropriate HBV status.

(A) Acute infection (incubation period)
(B) Acute infection (acute phase)
(C) Post infection/exposure
(D) Immunization
(E) HBV carrier state

491. IgG antibodies to core antigen, antibodies to E antigen, antibodies to surface antigen

492. HBsAg positive, HBeAg positive

493. HBsAg positive, HBeAg positive, IgM core antibody positive

494. HBsAg positive, no antibodies to HBsAg, other tests variable

495. Antibodies to HBsAg

Immunology

Questions 496–500

There are a variety of immunologic tests available for the detection of both antigen and antibody. Match the summaries of characteristics with the correct test.

(A) Latex agglutination (LA)
(B) Enzyme-linked immunosorbent assay (ELISA)
(C) Enzyme multiplied immunoassay test (EMIT)
(D) Counterimmunoelectrophoresis (CIE)
(E) Coagglutination (COA)

496. Combines features of gel diffusion and immunoelectrophoresis; applicable to only negatively charged antigens

497. Depends on the presence of protein A on certain strains of *Staphylococcus aureus*

498. Homogenous immunoassay; preferred for detection of low-molecular-weight substances

499. Used extensively to detect microbial antigens rapidly (5 min or less); inert particles are sensitized with either antigen or antibody

500. Heterogenous immunoassay; detection system is based on enzymatic activity

Immunology
Answers

414. The answer is C. *(Mandell, 3/e. p 1547.)* Bruton's agammaglobulinemia is a congenital defect that becomes apparent at approximately 6 months of age when maternal IgG is diminished. The child is unable to produce immunoglobulins and develops a series of bacterial infections characterized by recurrences and progression to more serious infections such as septicemia. Cell-mediated immunity is not affected and the child is able to respond normally to diseases that require this immune response for resolution.

415. The answer is C. *(Bellanti, 3/e. pp 101–103.)* Following is a schematic figure of Ig. Note that each peptide chain is drawn as a continuous line and attachments between heavy and light chains are noted by solid bars. There appears to be considerable flexibility in the hinge region (indicated on diagram) between the Fc and the two Fab portions of the molecule. This allows the molecule to assume either a T shape (in the diagram) or a Y shape.

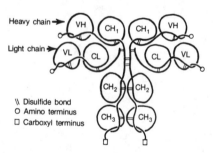

416. The answer is E. *(Davis, 4/e. pp 445, 472.)* Identification of NK cells is difficult because they can display either B- or T-cell characteristics. Some will form weak rosettes with sheep red blood cells (SRBCs) but unlike T cells have only a weak affinity for SRBCs. They appear not to have immunoglobulin receptors or I-A proteins like B cells. Activity is spontaneous to neoplastic cells and virally infected cells. Prior antigen exposure is not necessary—therefore the name "natural" killer cell.

417. The answer is C. *(Davis, 4/e. pp 406–407.)* Anaphylactic reactions occur within 3 to 4 months after antigen presentation. The response is attributed to IgE immunoglobulins that bind with high affinity to tissue mast cells and blood basophils. If the antigen is injected intravenously, systemic anaphy-

Immunology Answers

laxis—which can lead to shock, vascular engorgement, and asphyxia—occurs.

418. The answer is C. *(Bellanti, 3/e. pp 452–453.)* Bence Jones proteins are homogeneous free-globulin light chains (kappa and lambda chains) present in the urine of about half of all persons with multiple myeloma. Because Bence Jones protein is not albumin, "dipstick" reagents often employed to monitor urine for "protein" (i.e., albumin) are ineffective. Mu and gamma chains are types of heavy chains.

419. The answer is B. *(Jawetz, 19/e. pp 124, 125.)* Whether an antigen will induce tolerance rather than an immunologic response is largely determined by immunologic maturity of the host, structure and dose of antigen, and administration of immunosuppressive drugs. Tolerance is best maintained by the presence of low concentrations of antigen T cells become tolerant more readily than B cells.

420. The answer is D. *(Jawetz, 19/e. pp 108–110.)* Each secretory IgA molecule has an MW of 400,000 and consists of two H_2L_2 units and one molecule each of J chain and secretory component. Some IgA exists in serum as a monomer H_2L_2 with an MW of 170,000. Some bacteria, such as *Neisseria*, can destroy IgA1 by producing protease. It is the major immunoglobulin in milk, saliva, and tears.

421. The answer is B. *(Davis, 4/e. pp 476–477.)* At onset of symptomatic HIV disease, the number of helper T cells (CD4+) decreases. The number of suppressor T cells (CD8+) remains normal. This causes the normal ratio of CD4+/CD8+ cells to decrease from 2/1 to less than 1/1. In AIDS patients, a marked increase in serum immunoglobulins is usually seen.

422. The answer is E. *(Davis, 4/e. p 478.)* Except squamous epithelial cells, all the cells listed exhibit CD4 glycoprotein on their cell surface, although at lower levels than do helper T cells. This explains in part how HIV infection may be neurotropic. Because of the low levels of CD4 glycoprotein, these cells, as well as colon epithelial cells, are susceptible to HIV infection.

423. The answer is D. *(Davis, 4/e. pp 869, 893–894.)* Recurrent severe infection is an indication for clinical evaluation of immune status. Live vaccines, including BCG attenuated from *Mycobacterium tuberculosis*, should *not* be used in the evaluation of a patient's immune competence because patients with severe immunodeficiencies may develop an overwhelming infection from the vaccine. For the same reason, oral (Sabin) polio vaccine is not advisable for use in such persons.

168   **Microbiology**

424. The answer is E. *(Jawetz, 19/e. pp 117, 143.)* Interleukin 1 is a protein produced by macrophages that has three biologically active forms: IL-1α, β, and γ. Its functions include activation of B cells and stimulation of helper and cytotoxic T cells. Its activity is not histocompatibility-restricted.

425. The answer is C. *(Wilson, 12/e. pp 689–692.)* Chronic mononucleosis is an identifiable syndrome characterized by chronic malaise, often following an emotional stress. Unfortunately, chronic mononucleosis has become a catch-all diagnosis for many other infectious and noninfectious diseases that cause similar symptoms. The EBV pattern usually seen in chronic infectious mononucleosis is characterized by an elevated VCA-IgG, no VCA-IgM, low-titer EBNA, and elevated EA (1:20).

426. The answer is C. *(Davis, 4/e. p 437.)* Interleukin 1 (IL-1) is produced by a variety of cells. Primarily it is produced by activated macrophages or monocytes, although it can also be produced by activated B cells, keratinocytes, skin layer hairy cells, and natural killer cells. IL-1 acts synergistically with IL-6 to stimulate production of IL-2. Human IL-1 consists of α and β forms, or two different proteins. They have limited similarity of amino acid sequence and the β form is more abundant as a serum protein.

427. The answer is B. *(Davis, 4/e. p 285.)* As the following diagram illustrates, the hinge region of IgG exists between C_{H1} and C_{H2}.

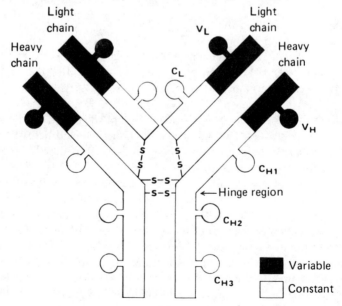

C regions are polypeptide segments of constant sequence; V regions have a variable amino-acid sequence. They occur in both the heavy (H) and light (L) chains. Differences in polypeptide sequence undoubtedly contribute to the distinctive biologic properties of the various immunoglobulins.

428. The answer is B. *(Davis, 4/e. pp 421–422.)* The Arthus reaction is a classic inflammatory response involving a cellular infiltrate provoked by antigen and antibody in much larger quantities than those required for passive cutaneous anaphylaxis. The edema of cutaneous anaphylaxis appears within 10 min and resolves within 30 min of antigen injection, but the polymorphonuclear leukocyte infiltrate of an Arthus reaction appears after more than an hour, peaks at 3 to 4 h, and resolves within 12 h. The severity of the Arthus reaction is proportional to the amount of antigen and antibody reacting. With high antigen-antibody concentrations, necrosis may result.

429. The answer is D. *(Davis, 4/e. pp 466–467.)* Allograft rejection is primarily a T-cell response to foreign tissue. The corticosteroids reduce inflammatory response and are generally administered by cytotoxic drugs, such as cyclosporine. Lymphoid irradiation is usually done so that the bone marrow is shielded. This removes lymphocytes from lymph nodes and spleen while allowing the patient to have the capacity to regenerate new T and B cells. Likewise, antilymphocyte globulin will destroy the recipient's lymphocytes, especially the T cells. Destruction of donor B cells would not play a role in the immunosuppression of the graft recipient. In graft crises, monoclonal antibody to CD3 is sometimes given. This targets mature T lymphocytes for destruction.

430. The answer is D. *(Mandell, 3/e. pp 1034–1043.)* Measleslike virus has been isolated from the brain cells of patients with subacute sclerosing panencephalitis (SSPE). The role of the host immune response in the causation of SSPE has been supported by several findings including the following: (1) progression of disease despite high levels of humoral antibody; (2) presence of a factor that blocks lymphocyte-mediated immunity to SSPE-measles virus in SSPE cerebrospinal fluid (CSF); (3) lysis of brain cells from SSPE patients by SSPE serum or CSF in the presence of complement (a similar mechanism could cause in vivo tissue injury).

Higher-than-normal levels of serum antibodies (Ab) to measles virus and local synthesis of measles Ab in CSF, as evidenced by the oligoclonal IgG, imply a connection between the virus and multiple sclerosis. However, other studies have implicated other viruses. Several studies of cell-mediated hypersensitivity to measles and other viruses in MS have been done, but the results have been conflicting. Definite conclusions regarding defects in cellular immunity in this disease cannot be reached until further research is completed.

431. The answer is A. *(Davis, 4/e. pp 283, 288–290.)* The serum concentrations (mg/mL) for each of the subclasses of immunoglobulins listed in the question are as follows: IgG-1, 8; IgG-2, 4; IgG-3, 1; IgG-4, 0.4; and IgA-1, 3.5. Serum concentrations of other antibodies are as follows: IgA-2, 0.4; IgM, 1; IgD, 0.03; and IgE, 0.0001. Subclasses of IgG differ in several respects, including effector function and, to a lesser degree, antigenic characteristics.

432. The answer is B. *(Davis, 4/e. pp 463–465.)* In the question presented, the haplotypes of the father are 3,25 and 7,12 and the haplotypes of the mother are 1,3 and 8,9. (A haplotype is composed of one allele—antigen—from one gene of a pair and one allele from the other gene.) Each child of this couple would have inherited one haplotype from each parent. Thus, possible offspring haplotypes are (i) 3,25; 1,3; (ii) 3,25; 8,9; (iii) 7,12; 1,3; and (iv) 7,12; 8,9.

433. The answer is C. *(Davis, 4/e. pp 436–437.)* While macrophages may be activated by CD4+ T cells, microbial products such as LPS and muramyl dipeptide may also serve as macrophage activators. Activated macrophages express MHC-II proteins on their surface and serve to present antigen to CD4+ T cells. They also secrete a variety of substances including IL-1, tissue necrosis factor, lysosomal enzymes, and reactive oxygen intermediates.

434. The answer is D. *(Davis, 4/e. pp 256–258.)* In tube D in the question presented, the maximum protein precipitate is observed. According to the rules governing precipitin reactions, maximum precipitation occurs at approximately antigen-antibody equivalence. In tubes A through C, antibody excess occurs; in tube E, antigen excess occurs.

435. The answer is D. *(Davis, 4/e. pp 379–383.)* The graph shown in the question exhibits hemagglutinating antibody responses to primary and secondary immunization with any standard antigen. Curve B represents the early response to primary immunization, which is chiefly an IgM response. Rechallenge elicits an accelerated response that mainly involves IgG and occurs 2 to 5 days after reimmunization. IgM has a molecular weight of 900,000 and is a pentamer that the fetus can produce quite early in gestation.

436. The answer is B. *(Davis, 4/e. pp 256–258.)* The ascending limb A of the precipitin curve presented in the question represents the zone of antibody excess; in this zone, the supernatant solution would contain unreacted antibody. On the descending limb C, or the antigen-excess zone, the supernatant solution contains excess free antigen. In a monospecific system, B designates the region of maximum precipitation and the supernatant solution is free to precipitate antibody and antigen. In a complex multispecific system, excess antigen or antibody molecules may be present at the point of maximum precipitate formation because the optimal quantity of each antigen may be different.

437. The answer is A. *(Davis, 4/e. pp 463–465.)* From the point of view of ABO compatibility, an A− person would not be expected to reject either type A or type O transfused platelets. However, an Rh− person should not receive Rh+ blood. Thus, the donor who would be HLA-compatible, as well as ABO- and Rh-compatible, would be O−, 1,3; 7,12.

438. The answer is A. *(Jawetz, 19/e. pp 115–116.)* Mouse T-cell markers include Ia, QA1, and Ly1. Equivalent markers in humans are HLA-DR and CD4. Originally, markers were defined by the company that produced the antisera, e.g., "T" by Ortho and "Leu" by Becton Dickinson. To clarify nomenclature, T-cell markers are now referred to as "CD" for *cell differentiation*. Immunoglobulins are found on the surface of B cells, not T cells.

439. The answer is A. *(Davis, 4/e. pp 391–394.)* Both IgG and IgM activate complement by the classic pathway, while IgA activates it by the alternative pathway. Neither IgD nor IgE can activate complement and, in fact, these immunoglobulins are short-lived in the human host, where they have a half-life of only 2.8 and 1.5 days, respectively.

440. The answer is E. *(Davis, 4/e. pp 869, 893–894.)* With repeated immunization, higher titers of all antibodies are observed, and, as priming is repeated, the immune response recruits B cells of progressively greater affinity. The affinity of antibody for a hapten-protein complex rises, cross reactivity also rises, and the response becomes wider in specificity. As the number of antigenic sites detected per reacting particle increases, the avidity increases. In addition to shifts in the class of immunoglobulin synthesized in response to an antigen (IgM to IgG), shifts also may occur in the idiotype of antibody.

441. The answer is A. *(Davis, 4/e. pp 264–265.)* In the Ouchterlony agar-gel diffusion test, an antigen and a series of antibodies (or an antibody and a series of antigens) are allowed to diffuse toward each other. At the zone of optimal proportions of the reactants, a precipitin line occurs. Cross reactions between antigens or antibodies tested can be detected by (1) a shortening of the major precipitin band contiguous to the cross-reacting substance or (2) the identity of precipitin reaction between the two cross-reacting substances. The figures presented in the question illustrate both types of cross reaction. In the first bleed pattern shown in the question, cross reaction between antigen X and antigen A is recognizable only by the shortening of the precipitin band between the center well and X on the A well side (relative to the band going directly into the normal saline well). In the second bleed pattern, full cross reaction of X and A is apparent. No other cross reactions are seen.

442. The answer is C. *(Davis, 4/e. pp 249–252.)* In a Scatchard plot, the slope of the line is equal to $-K$. As shown in the graph presented with the question, the slope ($r/c \div r$) is $-[(2 \times 10^4 \text{ liters/mole}) \div 2] = -1 \times 10^4$ liters/mole. Thus, K equals 1×10^4 liters/mole.

443. The answer is D. *(Davis, 4/e. pp 249–252.)* In the graph presented with the question, as r approaches 4, r/c approaches 0, and, consequently, c approaches infinity. In general, the x-intercept (that is, $r/c = 0$) gives the number of ligand binding sites at maximal saturation. In the example described, this maximum number of ligand molecules able to be bound per antibody molecule—or the antibody valence—is 4. Antibody valence also can be calculated from the equation given; that is, if $r/c = 2 \times 10^4$ liters/mole, $r = 2$, and $K = 1 \times 10^4$ liters/mole, then $n = [(r/c + Kr) \div K] = [(2 \times 10^4 \text{ liters/mole} + 2 \times 10^4 \text{ liters/mole}) \div 1 \times 10^4 \text{ liters/mole}] = 4$.

444. The answer is A. *(Davis, 4/e. pp 249–252.)* Secretory IgA is a tetravalent dimer and thus would have an antibody valence of 4 (as calculated in the previous question). IgG and IgE are divalent immunoglobulins. IgM is pentavalent or decavalent, depending on the experimental conditions.

445. The answer is E. *(Davis, 4/e. pp 319–341.)* Transfer factor, derived from human leukocytes, has been found to transfer cell-mediated responses in nonreactive recipients within 1 to 7 days after injection. Characterization of this substance (molecular weight less than 10,000) has been difficult; however, it is resistant to nucleases and proteases, is inactivated by heat (56°C, 30 min), and may be clinically effective in patients with cell-mediated immune deficiency.

446. The answer is D. *(Davis, 4/e. pp 256–261.)* Neither monovalent antigen nor monovalent antibody (Fab′) can form a precipitin lattice. An antigen molecule containing closely repeating antigenic determinants (e.g., a polysaccharide or a multichained polymeric protein) can bind antibody to two determinants on a single particle; this "monogamous bivalency" inhibits precipitation. F(ab′)$_2$ divalent antibodies can precipitate antigens, though they lack Fc portions.

447–451. The answers are: 447-B, 448-A, 449-B, 450-D, 451-A. *(Davis, 4/e. pp 478–481.)* T-cell function may be measured by skin testing or by lymphocyte proliferation. In the skin test, antigen is injected intradermally and examined for induration after 48 h. The antigens chosen are usually commonly encountered ones, such as *Candida* or streptokinase-streptodornase from hemolytic streptococci.

Serum immunoglobulin levels measure B-cell function and such measurement is usually done by radial immunodiffusion. In a normal population, immunoglobulin levels will vary, but levels less than 2 mg/mL for IgG and less than 10 μg/mL for IgA are considered deficient. Biopsy of lymph nodes that drain intracutaneous sites of antigen injection may be examined for plasma cells. If there are few or no plasma cells seen, this is an indication of B-cell dysfunction.

White blood cell counts typically enumerate the total number of lymphocytes per milliliter of blood when a differential is done. This will not separate T cells from B cells and therefore is not an effective method of separating T- and B-cell disorders.

452–456. The answers are: 452-D, 453-B, 454-E, 455-A, 456-C. *(Davis, 4/e. pp 269, 270–271, 275, 276.)* Affinity chromatography is used to purify antibodies and usually employs a column of agarose beads to which antigen has been attached. The serum or immunoglobulin solution is passed through the column so that the antibody can selectively attach to the antigen. All extraneous material can then be washed away. Finally, the desired product can be eluted.

Radial immunodiffusion is used to measure antigen or antibody concentrations. Agar is seeded with uniformly dispersed antisera and the antigen is placed in a well. A ring of precipitation develops where antigen-antibody reaction occurs. The area of the ring of precipitation is proportional to the amount of antigen.

Rocket immunoelectrophoresis is a combination of radial immunodiffusion and electrophoresis. It is capable of providing a more rapid method of quantitation than is radial diffusion. Antigen is applied to a well in agar that contains antisera. An electric current is used to shorten the time necessary for maximum precipitation. The height of the rocket-shaped zone of precipitation is proportional to the antigen concentration.

Western blot involves the transfer of protein from a polyacrylamide gel matrix to nitrocellulose. Bacterial or viral proteins are separated by molecular weight through electrophoresis. The protein fingerprint is then transferred to nitrocellulose. The nitrocellulose acts as an antigen matrix, which then interacts with sera. Specific antibodies can be visualized by either radioactive or enzymatic methods.

Enzyme-linked immunosorbent assay (ELISA) employs covalently linked enzyme-antibody complexes to demonstrate antigen-antibody interactions. A solid phase is coated with a specific antigen. A patient sample that contains antibody is added to the solid phase. Specific antibody is captured and free (i.e., nonspecific) antibody is removed by washing. The presence of the antigen-antibody sandwich is detected by the addition of enzyme-labeled antihuman antibody. The enzyme in the presence of its substrate will produce a detectable reaction (usually colored, although it may be fluorescent). Thus, the addition of substrate allows visualization of the enzyme-labeled antigen-antibody complex. Reactions may be read visually or spectrophotometrically or by fluorometry.

457–461. The answers are: 457-E, 458-A, 459-B, 460-D, 461-C. *(Grieco, pp 295–299, 303–307, 311–312, 316.)* Borrelia burgdorferi, the causative agent of Lyme disease, elicits an acute antibody response. IgM appears within days to a few weeks following a tick bite and IgG appears a few weeks later. IgG persists, IgM does not. Cross reactions occur with other treponemes.

Fifth disease is a viral exanthem commonly seen in children 8 to 12 years old. Children are ill for a few days but recover without incident. Unfortunately, if a pregnant female acquires the disease in the first trimester of pregnancy, the fetus is at risk. The causative agent is thought to be a parvovirus (parvovirus B19).

Adults with no titer to varicella (VZV) are at risk for acquisition of chickenpox. If they are health care workers, there is additional risk in transmitting VZV to immunodeficient children. Antibodies to VZV are readily detected by both EIA and FA techniques.

Delta agent is a recently discovered antigen associated with HBsAg. Its presence usually correlates with HBsAg chronic carriers who have chronic active hepatitis. EIA and RIA tests are available to detect antibodies to delta agent.

SSPE is thought to be caused by a measles-related virus present in the central nervous system. Most SSPE patients show elevated measles virus antibodies in serum and CSF. In patients with multiple sclerosis (MS), lower CSF antibody titers have been observed, suggesting a possible etiologic role for measles virus in MS.

462–465. The answers are: 462-D, 463-B, 464-C, 465-E. *(Mandell, 3/e. pp 1178–1180.)* Epstein-Barr virus (EBV), a member of the human herpesvirus group, has been established as the causal agent of heterophil-positive and heterophil-infectious mononucleosis. The diagnosis of infectious mononucleosis can be made in 80 to 90 percent of cases by demonstration of heterophil antibodies. In the heterophil-negative cases and for atypical infections, determination of specific antibodies to EBV are useful in establishing the diagnosis.

The development of IgM antibodies to the viral capsid antigen (EBV-

VCA [IgM] Ab) can be used for specific diagnosis of a current or recent infection. They arise early in the disease and persist for only 4 to 8 weeks. On the other hand, many adults have been exposed to EBV and maintain a low level of IgG in response to this virus. A titer of greater than 1 to 640 is usually diagnostic.

Antibodies to the early antigen of EBV (EA-EBV) show two distinct patterns of fluorescence—diffuse staining of both nuclei and cytoplasm (anti-D) and staining of cytoplasmic aggregates (anti-R). Anti-D titers appear 3 to 4 weeks after onset and are present in 70 percent of EBV-induced mononucleosis. Titers correlate with the severity of clinical illness and disappear after recovery. Anti-R titers appear 2 weeks to several months after onset and are rarely seen in cases of infectious mononucleosis. They are detectable in atypical cases and remain so for up to 2 years.

Antibodies to the nuclear antigen of EBV (EBNA-Ab) appear 3 to 4 weeks after onset and persist for life. They are useful in assessing recent infection if patients are VCA-positive, EBNA-negative and then become EBNA-positive.

466–470. The answers are: 466-D, 467-C, 468-D, 469-B, 470-D. *(Davis, 4/e. pp 478–481.)* Immunodeficiency disorders can be categorized according to whether the defect primarily involves humoral immunity (bone marrow–derived, or B, lymphocytes) or cellular immunity (thymus-derived, or T, lymphocytes) or both. Swiss-type hypogammaglobulinemia, ataxia-telangiectasia, the Wiskott-Aldrich syndrome, and severe combined immunodeficiency disorders all involve defective B-cell and T-cell function. Infantile X-linked agammaglobulinemia is caused chiefly by deficient B-cell activity, whereas thymic hypoplasia is mainly a T-cell immunodeficiency disorder.

471–474. The answers are: 471-D, 472-C, 473-A, 474-B. *(Davis, 4/e. p 463.)* Transplantation from one region of a person to another region of that same person is an autograft and has the best chance of succeeding. When a transplant is done between monozygotic twins, it is an isograft and has a complete MHC compatibility and a good chance of success. Allografts are between members of the same species and xenografts are between members of different species. Both of these transplants have a high rate of rejection unless immunosuppression accompanies the transplant.

475–477. The answers are: 475-D, 476-C, 477-B. *(Davis, 4/e. pp 277–298.)* Isotypes are determined by antigens of the major immunoglobulin classes found in all individuals of one species. In addition to heavy-chain isotypes of IgA, IgD, IgE, IgG, and IgM, two light-chain isotypes exist for kappa and lambda chains.

Allotypes are differentiated by antigenic determinants that vary among individuals within a species and are recognized by cross-immunization of individuals in a species. Allotypes include the Gm marker of IgG and the Inv marker of light chains.

Idiotypes are antigenic determinants that appear only on the Fab fragments of antibodies and appear to be localized at the ligand-binding site; thus, anti-idiotype antisera may block reactions with the appropriate hapten. The carbohydrate side chains of immunoglobulins are relatively nonimmunogenic. New determinants may be exposed after papain cleavage of immunoglobulins, but these determinants are not included in the classification of the native molecule.

478–482. The answers are: 478-B, 479-C, 480-D, 481-E, 482-A. *(Davis, 4/e. pp 278–282.)* Electrophoresis of human serum proteins identifies five distinct types: albumin, α_1-proteins, α_2-proteins, β-proteins, and gamma globulins. A normal electrophoretic profile appears below:

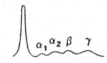

Many human diseases can be diagnosed, at least in part, on the basis of abnormal electrophoretic profiles. For example, absence of the second peak (α_1) is compatible with a diagnosis of α_1-antitrypsin deficiency in symptomatic persons. A sharp and high gamma peak indicates the presence of a monoclonal gammopathy, such as multiple myeloma; on the other hand, a gamma peak that is diffusely elevated points to a polyclonal hypergammaglobulinemia. Complete absence of the gamma peak is associated with Swiss-type agammaglobulinemia.

483–485. The answers are: 483-C, 484-B, 485-B. *(Jawetz, 19/e. pp 121–122.)* The complement-fixation (CF) test is a two-stage test. The first stage involves the union of antigen with its specific antibody, followed by the fixation of complement to the antigen-antibody structure. In order to determine whether complement has been "fixed," an indicator system must be employed to determine the presence of free complement. Free complement binds to the complexes formed when red blood cells (RBCs) are mixed with anti-RBC antibody; this binding causes lysis of the cells. Complement that has been "fixed" before addition of red blood cells and anti-RBC antibody cannot cause lysis.

Immunology Answers

486–490. The answers are: 486-E, 487-C, 488-B, 489-A, 490-D. *(Grieco, pp 234–240.)* Infectious mononucleosis (IM) may be suspected clinically but it is confirmed serologically. The heterophil antibody test, however, may be negative in up to 15 percent of adults and 35 to 40 percent of children. Because IM is caused by Epstein-Barr virus (EBV), a specific test for the viral capsid antigen (VCA) of EBV is indicated when heterophil tests are negative. Because a portion of the normal population has IgM antibodies to EBV, IgM antibodies to VCA may be necessary.

Mycoplasma pneumoniae causes rather severe primary atypical pneumonia. Although physicians have relied on the presence of cold agglutinins (CA) for diagnosis, CA may be negative in up to 50 percent of cases. A complement-fixation (CF) test for *M. pneumoniae* is indicated when CA is negative. Isolation of *M. pneumoniae* is time-consuming (1 to 3 weeks) and not practical in most circumstances.

Rubella immune status tests are usually done by hemagglutination inhibition (HI), enzyme immunoassay (EIA), or latex agglutination. IgM tests for rubella are often necessary in congenital infection in order to separate maternal from fetal antibodies.

The VDRL test for syphilis has been supplanted in many laboratories by the rapid plasma reagin (RPR) test. In the RPR test the patient's serum is mixed on a card with RPR antigen and sensitized charcoal particles. If antibody is present the particles clump. The RPR is more sensitive than the VDRL, but, like the VDRL, false positives may occur in 15 to 20 percent of the positive RPR results.

Diagnosis of disseminated staphylococcal infection or staphylococcal endocarditis may be difficult, particularly in partially treated patients. Detection of teichoic acid antibodies to the ribitol teichoic acid in the cell wall of *S. aureus* may be helpful. Because other staphylococci as well as some streptococci also have teichoic acid, the test may not be as specific as desired. In any event, a single teichoic acid Ab titer is rarely useful and a fourfold rise in titer is necessary for diagnosis.

491–495. The answers are: 491-C, 492-A, 493-B, 494-E, 495-D. *(Grieco, pp 295–299.)* The diagnosis of HBV infection is usually based on three tests: hepatitis B surface antigen, antibodies to surface antigen, and antibodies to core antigen. Tests are available, however, for e antigen and antibodies to e antigen. A variety of testing methods are available and include enzyme immunoassay, radioimmunoassay, hemagglutination, latex agglutination, and immune adherence. The delta agent has recently been described. The delta agent exacerbates infecton with HBV, apparently in a synergistic manner. Commercial tests are now available for delta agent. The table that follows presents the patterns of hepatitis B virus serologic markers observed in various stages of infection with HBV.

	Serologic Markers					
Interpretation	HBsAg	HBeAg	IgM Anti-HBc	Total Anti-HBc	Anti-HBe	Anti-HBs
Acute infection						
Incubation period	+*	+*	–	–	–	–
Acute phase	+	+	+	+	–	–
Early convalescent phase	+	–	+	+	+	–
Convalescent phase	–	–	+	+	+	–
Late convalescent phase	–	–	–†	+	+	+
Long past infection	–	–	–	+‡	+ or –	+‡
Chronic infection						
Chronic active hepatitis	+§	+ or –	+ or –	+§	+ or –	–§
Chronic persistent hepatitis	+¶	+ or –	+ or –	+	+ or –	–
Chronic HBV carrier state	+¶	+ or –	+ or –	+	+ or –	–
HBsAg immunization	–	–	–	–	–	+

*HBsAg and HBeAg are occasionally undetectable in acute HBV infection.

†IgM anti-HBc may persist for over a year after acute infection when very sensitive assays are employed.

‡Total anti-HBc and anti-HBs may be detected together or separately long after acute infection.

§HBsAG-negative chronic active hepatitis may occur where total anti-HBc and anti-HBs may be detected together, separately, or not at all.

¶HBsAg-negative chronic persistent hepatitis and chronic HBV carriers have been observed.

496–500. The answers are: 496-D, 497-E, 498-C, 499-A, 500-B. *(Grieco, pp 224–232.)* Of the many methods available for antigen and antibody detection, LA, ELISA, EMIT, CIE, and COA are the most widely used.

Latex agglutination (LA) employs latex polystyrene particles sensitized by either antibody or antigen. LA is more sensitive than CIE and COA but slightly less sensitive than either RIA or EIA. LA has been used to detect *Haemophilus influenzae, Neisseria meningitidis,* and *Streptococcus pneumoniae* antigens in cerebrospinal fluid. LA has also been used for detection of cryptococcal antigen. Most recently, LA has been widely used for rapid detection of group A streptococcal antigen directly from the pharynx. The test is rapid (5 min), sensitive (approximately 90 percent), and specific (99 percent).

Coagglutination (COA), also an agglutination test, is slightly less sensitive than LA but less susceptible to changes in environment (e.g., temperature). Most strains of coagulase-positive staphylococci have protein A in their cell wall. Protein A binds the Fc fragment of antibody immunoglobulin, leaving the Fab portions to capture homologous antigen. Similar to LA, COA has been widely used to detect microbial antigens in body fluids. COA has also been used to rapidly type or group bacterial isolates.

Enzyme immunoassays (EIAs) can be either homogenous (EMIT) or heterogenous (ELISA). EMIT has been used primarily for assays of low-molecular-weight drugs. Its primary use in microbiology has been for assays of aminoglycoside antibiotics. EIAs vary as to the solid support used. A variety of supports can be used, such as polystyrene microdilution plates, paddles, plastic beads, and tubes. The number of layers in the antibody-antigen sandwich varies; usually as additional layers are added, detection sensitivity is increased. The two most common enzymes are horseradish peroxidase (HRP) and alkaline phosphatase (AP). Beta galactosidase has also been employed. Orthophenylene diamine is the most common substrate for HRP and p-nitrophenyl phosphate for AP. Because EIAs are usually read in the visible color range, the tests can be read qualitatively by eye or quantitatively by machine.

Counterimmunoelectrophoresis (CIE) was originally used for "Australia antigen" (HBsAg) but was soon replaced by RIA. For a decade, CIE was used to detect antigens in body fluids. CIE is not an easy technique. Its success depends on the control of many variables, including solid support, voltage, current, buffer, affinity and avidity of antibodies, charge on the antigen, and time of electrophoresing.

Bibliography

Ash LR, Orihel TC: *Atlas of Human Parasitology,* 3/e. Chicago, ASCP, 1990.

Balows A, Hausler WJ, Lennette EH: *Laboratory Diagnosis of Infectious Diseases: Principles and Practice,* vols 1 and 2. New York, Springer-Verlag, 1988.

Balows A, et al: *Manual of Clinical Microbiology,* 5/e. Washington, DC, American Society for Microbiology, 1991.

Baron S: *Medical Microbiology,* 3/e. New York, Churchill Livingstone, 1991.

Bellanti JA: *Immunology in Medicine,* 3/e. Philadelphia, WB Saunders, 1985.

Davis BD, et al: *Microbiology,* 4/e. New York, Harper & Row, 1990.

Grieco M, Meriney DK: *Immunodiagnosis for Clinicians: Interpretation of Immunoassays.* Chicago, Year Book Medical, 1983.

Howard BJ, Klass J, Rubin SJ, Weissfeld AS, Tilton RC: *Clinical and Pathogenic Microbiology.* St. Louis, CV Mosby, 1987.

Jawetz E, Melnick JL, Adelbert EA: *Review of Medical Microbiology,* 19/e. East Norwalk, CT, Appleton & Lange, 1991.

Lorian V (ed): *Significance of Medical Microbiology in the Care of Patients.* Baltimore, Williams & Wilkins, 1986.

Mandell GL, Douglas RG, Bennett JE: *Principles and Practice of Infectious Disease,* 3/e. New York, Wiley & Sons, 1990.

MMWR 37:737–738, 1988.

MMWR 38:S–7, 1989.

Volk WA, et al: *Essentials of Medical Microbiology,* 3/e. Philadelphia, JB Lippincott, 1986.

Wilson JD, et al: *Harrison's Principles of Internal Medicine,* 12/e. New York, McGraw-Hill, 1991.